DFPC SERIES power cables
> Skin-filtering technology turns these power cords into a serious component upgrade.

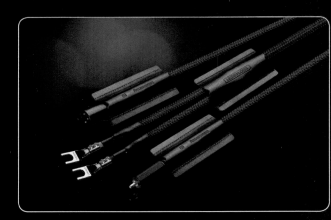

ANCHORWAVE IC and speaker cable
> High-tech cloth shielding and an amorphous nanoparticle bed ensure high performance.

FIREWALL power conditioning unit
> Skin-filtering technology and micro-vibration damping panzerholz work in tandem.

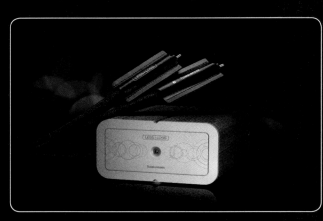

TUNNELBRIDGE distortionless IC system
> Via two separate paths, Tunnel and Bridge, distortionless signal transfer is achieved.

BLACKBODY ambient field conditioner
> Near ideal absorption and reflection pattern pushes the boundaries to new frontiers.

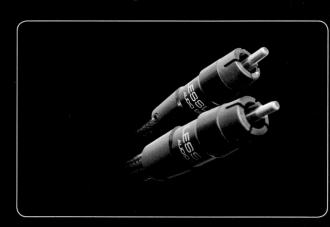

75 / 110 OHM DIGITAL cables
> A reflection-attenuation network and 5 shielding methods ensure low jitter transfer.

Audio Annual

Issue 1 | 2014

Audio Annual

Issue 1 | 2014

Editorial & Production

Mathew Thalakotur

Surya Thalakotur

Design

Surya Thalakotur

Studio Equinox

Contact & Submissions

mathew@thalakotur.com

 @audioannual

 @audioannual

www.audioannual.com

© 2013 Thalakotur LLC. All Rights Reserved.

Reproduction and/or distribution of this publication, in whole or in part, is forbidden without prior written permission from Thalakotur LLC. The information contained herein has been obtained from sources believed to be reliable; nevertheless, Thalakotur LLC makes no representations or warranties as to the accuracy, completeness or adequacy of such information. The opinions shared are those of the contributing authors and not necessarily reflective of Thalakotur LLC and/or its affiliates. Thalakotur LLC shall not be liable for any errors, omissions and/or inadequacies in the information contained herein or for interpretations thereof. The opinions expressed herein are subject to change without notice.

The Definitive Audiophile Reference.

The audiophile space is fragmented by its very nature. Companies of all sizes produce equipment around the world in pursuit of perfect sound reproduction, but not all of them are widely known. Our goal is to provide a window into that broader audiophile space. We seek to showcase innovation from the big names, as well as the smaller boutique manufacturers.

All the content in this magazine has been provided through our partnership with the manufacturers. Our role is to present the information needed to start your research, or simply to broaden your perspective on the category. With that in mind, we do not charge any of the featured manufacturers, allowing us to even the playing field.

Audio Annual is a private publication based in Cincinnati, Ohio. Though many magazines play in the high-end audio space, the industry lacked a single resource with a broad catalogue of manufacturers. Additionally, we decided against reviewing the products featured. Quite frankly, a quick Google search on any of these products or manufacturers will return feedback from a multitude of perspectives.

Thank you for purchasing Audio Annual. We hope you enjoy the content ahead.

Mathew Thalakotur

Mathew Thalakotur, Founder

08	**09**	**10**	**11**	**12**	**13**
Abbingdon Music Research United Kingdom	Acoustic Preference Slovenia	Albedo Audio Italy	Amphion Finland	Anthony Gallo Acoustics California, USA	Are Audio Canada
14	**15**	**16**	**17**	**18**	**19**
Audioengine Hong Kong	Bowers & Wilkins United Kingdom	Broadmann Austria	Earthquake Sound California, USA	Eclipse United Kingdom	Engelholm Audio Sweden
20	**21**	**22**	**23**	**24**	**25**
Eryk S Concept Poland	Everything But The Box Bulgaria	Ferguson Hill United Kingdom	Gato Audio Denmark	Genesis Advanced Technonology Washington, USA	German Physiks Germany
26	**27**	**28**	**29**	**30**	**31**
GoldenEar Technology Maryland, USA	Gradient Finland	Grado New York, USA	HHR Exotics Ohio, USA	Hoerning Hybrid Denmark	Kaiser Acoustics Germany
32	**33**	**34**	**35**	**36**	**37**
Kharma Netherlands	Kralk Audio United Kingdom	Legacy Illinois, USA	Mansberg Holland	Mark & Daniel Arizona, USA	Morel Israel
38	**39**	**40**	**41**	**42**	**43**
Mr. Speakers California, USA	Musique Concrete France	Mythos Audio Greece	Origin Live United Kingdom	Perfect Sound Taiwan	Perfect Sound Taiwan
44	**45**	**46**	**47**	**48**	**49**
Sanders Sound Systems Colorado, USA	Shape Audio Sweden	Shure Illinois, USA	Sonus Faber Italy	Sound & Design Italy	Soundlabs Utah, USA
50	**51**	**52**	**53**	**54**	**55**
Surreal Sound Virginia, USA	SVS Ohio, USA	Tetra Canada	TR Studio Poland	Waterfall Audio France	Westone Music Colorado, USA
56	**57**				
Wisdom Audio Nevada, USA	YG Acoustics Colorado, USA				

Abbingdon Music Research

United Kingdom
www.amr-audio.co.uk

Abbingdon Music Research (AMR) is a high-end manufacturer of some of the finest audio components. The 77 Reference Class, the 777 Premier Class and its range of accessories have all garnered critical acclaim. Since the CD-77, which was first launched in 2006, through to the latest DP-777 all boast ground-up technology. AMR's class re-defining audio components have justly forged an enviable reputation to move the listener with the stirring emotion, verve and faithfulness of the original performance.

LS-77
Desktop Speaker
$14,999 (pair)

The LS-77 does not sing, it soars: with a bewitching and true, full dynamic range that is free from coloration. AMR's professional monitor is able to: "Stir the listener into orchestrating to the grandest classical overture or shedding a tear at the end of an Operatic piece." The LS-77 boasts a substantial Magnesium-Aluminum alloy cabinet, custom-made isoplannar tweeter and material-matched proprietary 10" driver. Coupled with AMR's acoustically-inert epoxy-formed cherry plywood crossover, which is comprised of AMR's copper foil inductors and film foil capacitors terminated with the highest grade, dedicated Teflon signal wiring. The discerning listener will be taken aback by the workmanship and attention to detail in every aspect.

Drivers	10 inch and 4 inch Voice coil, 5 inch Isoplanar ribbon
Crossover	AMR Optislope design
Sensitivity	87dB
Peak Power Handling	600W
Power Handling	150W
Nominal Impedance	8 ohms
Frequency Response	26Hz – 40kHz
Minimum Impedance	6 ohms
Dimensions (W x H x D)	10 x 19 x 12 inches / 254 x 483 x 305 mm
Weight	77lbs / 35kgs
Available Finishes	Champagne, Titanium

Acoustic Preference

Slovenia
www.acoustic-preference.com

Acoustic Preference enterprise is a small family-run manufacturing company established in 2001. With the help of music, we can conjure up the majority of human feelings. At the same time, music represents a tool for relaxation and creativity, since it influences so many human senses. Acoustic Preference's goal is to create acoustic components that combine natural sounds and offer great delight when listening to music on every occasion. They want to present their clients with durability and a feeling of prestige and excellence.

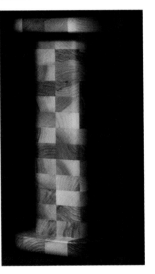

Gracioso 1.0
2-Way Loudspeaker
$11,403 (pair)

Gracioso is a speaker for those with the highest taste in performance and design. As it is aesthetically striking with its complex hardwood selected materials, so is the full embodied, warm, dynamic and transparent sound it produces. Gracioso takes the best components available and squeezes the last potential from them. Gracioso is all about music. You can't think about technical details when you want to get lost in a song and let yourself be personally invited into the audio performance. The final quality of any product will go only as far as the basic ingredients will allow.

Drivers	160mm Morel bass-midrange, 28mm Morel tweeter
Frequency Response	50Hz – 20kHz
Nominal Impedance	8 ohms
Sensitivity	88dB
Crossover	2.2kHz
Nominal Power	150W
Options	Stands with or without spikes
Dimensions (W x H x D)	10.2 x 14.1 x 14.3 inches / 260 x 360 x 365 mm
Weight	27.7lbs / 12.6kgs
Available Finishes	Red-Brown Europe Walnut, Dark Brown American Walnut

Albedo Audio

Italy
www.albedoaudio.com

In the early 90s the R&D department of Albedo faced the general issues of transmission line loudspeaker systems, formulating for the first time a rigorous mathematical model to describe this specific acoustic configuration. They found that a duct filled with fibrous stuff is losing in terms of efficiency and size and that the geometrical shape of the duct drastically affects the final results. Obviously, these basic principles were complemented with other completely original acoustic solutions like Helmholtz resonators used to equalize the duct emission.

Axcentia

3-Way Loudspeaker
$34,000 (pair)

The question is somewhat obvious: If small speakers like HL 2.2 and Aptica are able to give a wide and rich bass response, what would happen using a couple of larger drivers? The answer is in the Axcentia, where the Helmholine system, made by both wooden cabinet and metal base, shows all its power in managing high volume levels and in keeping the detailed bass texture which transmission lines are properly renowned for. The elliptically shaped cabinet is made in a special high density and well damped sandwich. All ceramic drivers, a sophisticated crossover and a metal chambered base are further important details.

Drivers	Ceramic - 2 x 6.5 inch woofers, 5 inch midrange, 1.2 inch tweeter
System	Linear Phase
Loading	Transmission line filtered by resonators
Crossover	Acoustic first order
Frequency Response	35Hz - 20kHz
Nominal Impedance	8 ohms
Sensitivity	89dB
Dimensions (W x H x D)	9.8 x 45.2 x 24.4 inches / 250 x 1150 x 620 mm
Weight	143.3lbs / 65kgs
Available Finishes	Makassar Ebony, Jamaica Ebony, Walnut
Grill	Black

Amphion

Finland
www.amphion.fi

Amphion Loudspeakers Ltd. was established in 1998. They design and build loudspeakers that are characterized by a high resolution, an unstrained low-level reproduction and an unrivalled voice band. Their reasonably priced solutions are rather insensitive to room acoustics and fill the room with a wide and even dispersion. A timeless yet individual approach with regard to aesthetics and application make Amphion products match with individual styles and preferences.

Argon7L

2-Way Loudspeaker
$4,995 (pair)

There are speakers that are classified as neutral. There are speakers that are classified as musical. Yet only a few can be claimed to possess "musical neutrality". Argon7L has ruler flat response. But it does not sound like that. Argon7L offers transparency and neutrality, but is never dry nor irritating. It offers high level of musicality, but is never colored. High quality drivers, extremely simple crossover and innovative enclosure structure provide a clean, wide window to all types of music. With high levels of transparency and neutrality, listeners can discover new thrilling details from their favorite recordings. Even the non-perfect ones.

Design	2-way, vented
Drivers	1 inch titanium, 2 x 6.5 inch aluminiun
Crossover Frequency	1600Hz
Nominal Impedance	4 ohms
Sensitivity	93dB
Frequency Response	28Hz – 30kHz
Power Recommendation	10W - 150W
Dimensions (W x H x D)	7.5 x 45.6 x 12 inches / 191 x 1160 x 305 mm
Weight	66.1lbs / 30kgs
Available Finishes	White, Black, Wood Cherry, Birch, Walnut
Grills	White, Black

Anthony Gallo Acoustics

California, USA
www.roundsound.com

Anthony Gallo was born in 1963. At around age 13, he became hooked on loudspeaker design and decided to pursue his dream of designing and manufacturing loudspeakers full-time. Anthony Gallo Acoustics was born. Joining soon thereafter with Don Fried, who provided key financial backing and took the reigns as CEO, Anthony Gallo Acoustics shifted its focus to the audiophile community and ascended toward becoming a major player in consumer electronics.

A'Diva SE
Loudspeaker
$329 (pair)

Introducing the A'Diva SE. The Special Edition utilizes a proprietary ultra-wide dispersion flat-diaphragm transducer that covers a frequency range from 80Hz to 22kHz, all powered by a single driver. This technology enables these spherical wonders to perform as true audiophile loudspeakers. Because it uses a single driver, the A'Diva SE has no need for a crossover. Whether used for 2-channel or Home Theater these compact satellites speakers fill the room with a vivid, clear 3-D sound stage that remains always stable.

Frequency Response	80Hz - 22kHz (on wall) , 100Hz - 22kHz (on stand)
Nominal Impedance	4 ohms
Sensitivity	85dB
Power Handling	60W (full range) , 125W (crossover at 80Hz -120Hz)
Driver	3 inch wide-dispersion flat diaphragm
Cone Material	Aluminum laminated proprietary honeycomb sandwich
Enclosure Material	Steel / Stainless Steel
Dimensions	5 inch sphere / 127 mm sphere
Weight	2lb 2oz / 0.96kgs
Available Finishes	Matte Black, Matte White, Stainless Steel
Grill	Black Cloth, White Cloth

Are Audio

Canada
www.areaudio.com

Are Audio was an endeavor initiated by Richie White and Ross Connolly, who have been best friends since they were kids. As they grew older they also shared a keen interest in engineering, physics, and hi-fi sound. Based out of St. John's Newfoundland, they created Are Audio. Ross continues hand-manufacturing the remarkable speakers that he and Richie designed, and currently operates the venture while still maintaining close ties with Richie, and still wiping out on bikes.

Light Roasts

2-Way Loudspeaker
$935 (pair)

The Light Roasts were devised over many many pots of coffee. With some math and a little background music, we had developed a simple enclosure based around a natural sounding woofer. Along with this driver, a quality silk dome tweeter of similar voicing, added with a clean first order network fit the equation. A few prototypes later and this simplistic design had won them over with its smooth midrange and great extension, and detailed highs. From two-channel stereo, to studios, to home theater, the Light Roasts can fill any role. Now, pour up a coffee and enjoy.

Type	2-ay reflex enclosure with top port
Drivers	26mm silk dome tweeter, 140mm cone with butyl rubber surround woofer
Frequency Response	50Hz – 20kHz
Nominal Impedance	4 ohms
Crossover	Point to point wired first order 3kHz
Power	60W
Dimensions (W x H x D)	6.4 x 12.9 x 10.2 inches / 165 x 330 x 260 mm
Weight	12.1lbs / 5.5kgs
Available Finishes	8 or more color choices

Audioengine

Hong Kong
www.audioengineusa.com

Audioengine products are based on custom designs with very few off-the-shelf parts. After years of building professional powered studio monitor speakers, Audioengine has taken their experience and created unique powered consumer speakers for your home and desktop. Now Audioengine is into it's 6th year and their goal remains the same - to give customers a high-end listening experience at affordable prices with easy to use, highquality products. Audioengine invites you to join the computer audio revolution!

Audioengine 5+

2-Way Bookshelf Speaker
$399 (pair)

Connect your iDevice, computer, TV, or any other audio component for great stereo sound in any room. The Audioengine 5+ Premium Powered Speaker System delivers the audio performance and aesthetic quality expected from Audioengine and continues to close the gap between your computer music and home hi-fi. Even if you're not an audio enthusiast, we guarantee your music will sound better!

Power Output	50W RMS / 75W peak per channel, AES
Inputs	1/8 inch stereo mini-jack, RCA L/R
Outputs	RCA L/R audio, USB Type A
Amplifier Type	Dual Class AB monolithic
Drivers	5 inch Kevlar woofers, 20 mm Silk dome tweeters
Signal-to-Noise Ratio	>95dB
Harmonic Distortion	<0.05%
Frequency Response	50Hz - 22kHz
Dimensions (W x H x D)	7 x 10.8 x 7.8 inches / 178 x 273 x 197 mm
Weight	15.4lbs / 6.9kgs (left); 9.6lbs / 4.3kgs (right)
Available Finishes	Satin Black, Hi-Gloss White, Solid Carbonized Bamboo

Bowers & Wilkins

England, UK
www.bowers-wilkins.com

In 1966, John Bowers and his lifelong friend Peter Hayward founded a manufacturing company, then called B&W Electronics, in Worthing, on the south coast of England. They agreed from the outset to live modestly and plough any profits back in to the business to further research into the quest for the perfect loudspeaker. Bowers & Wilkins loudspeakers are designed to reproduce sound as accurately as possible. Whether it's hi-fi, home theatre or custom installation, the audio performance of any Bowers & Wilkins loudspeaker is best in class.

CM10
3-Way Loudspeaker
$4,000 (pair)

The top-of-the-range CM10 stands out from the CM range crowd with its tweeter-on-top design. Isolating the tweeter aids imaging and dispersion for a more natural, spacious sound. The tweeter's unique "double dome" construction also prevents the voice coil from distorting at higher frequencies. The result raises the bar for precision and control even at the highest frequencies, but particularly in the audible spectrum. A third bass driver, in the CM10 delivers a powerful bass punch, but also reduced distortion and extended bass response.

Frequency Response	45Hz - 28kHz
Sensitivity	90dB
Harmonic Distortion	<1% 86Hz - 28kHz; <0.5% 110Hz - 20kHz
Crossover Frequency	350Hz, 4kHz
Recommended Amplification	30W - 300W
Nominal Impedance	8 ohms
Dimensions (W x H x D)	14.4 x 42.8 x 16.3 inches / 366 x 1087 x 414 mm
Weight	73.7lbs / 33.5kgs
Available Finishes	Real wood veneers: Rosenut, Wenge / Painted: Gloss Black, White
Grill	Black

Brodmann

Austria
www.brodmann.at

Every speaker system and every instrument that Brodmann produces is a direct reflection of their passion for music. They take great pride in tirelessly researching new technologies and procedures to improve products, which they consider their works of art. The company builds speakers not as mere vessels, but as instruments themselves. Brodmann has defined new standards for the authentic reproduction of music. Beautiful in performance and appearance, Brodmann speakers are destined to be the classics of tomorrow.

VC7
Loudspeaker
$24,990 (pair)

The Vienna Classic line is the fascinating result of landmark engineering, art in harmonic interplay with precision workmanship, and a technical solution fine tuned to the very last detail. Power and composure unite in a velvety soft and yet very detailed sound that will charm your senses in a way never before experienced from a mere speaker. Brodmann's premium VC 7 is a landmark in the hi-fi class, marking the pinnacle in building speaker systems. The VC 7 is their largest floor-standing speaker, and is created with one thing in mind: your absolute fascination when listening to music, whatever genre you choose.

Frequency Response	25Hz – 27kHz
Output Power	180W
Nominal Impedance	4 ohms
Transitional Frequency	Acoustic Active: 130Hz / Electrical: 1.6kHz
Sensitivity	91dB
Dimensions (W x H x D)	7.4 x 52.3 x 15.8 inches / 195 x 1330 x 403 mm
Weight	80.4lbs / 36.5kgs
Available Finishes	Black, Walnut, White and 7 others

Earthquake Sound

California, USA
www.earthquakesound.com

In 1984, Joseph J. Sahyoun, a music freak and Aerospace Engineer unhappy with the existing loud speaker technology and performance, decided to put his advanced engineering knowledge to use. He pushed technological boundaries and Earthquake quickly created a name for itself in the car audio industry. They became well known for their powerful subwoofers and amplifiers. Currently, Earthquake Sound can provide audio solutions in every room, whether in home theater, outdoors, or just in the background.

Titan Tigris

3-Way Loudspeaker
$9,500 (pair)

The Tigris like its baby brother Telesto descends from the Reference speaker Titan Tethys. The Tigris is a dominant and striking evolution of the uncompromising Titan series. Titan differentiates itself as everything is engineered from initial conception to work together. This improves all aspects of the design. The whole becomes greater than the sum of its parts. A speaker where detail is not achieved at the expense of dynamics. Titan Series straight through wide open performance sets a new standard for loudspeaker design.

Drivers	1 inch Silk dome, 2 inch Silk dome, 2 x 8 inch Kevlar woofers
Frequency Response	25Hz - 40kHz
Sensitivity	87dB
Nominal Impedance	4 ohms
Crossover Frequency	350Hz, 3.2kHz, 10kHz
Recommended Amplification	100W - 500W RMS
Dimensions (W x H x D)	10.8 x 51.3 x 19.2 inches / 274 x 1302 x 488 mm
Weight	89.3lbs / 40.5kgs
Available Finish	Fine high gloss 3mm black lacquer paint

Eclipse

United Kingdom
www.eclipse-td.net

High-end speaker specialist Eclipse TD is the "Formula 1" division of the huge Fujitsu-Ten audio group, which gives Eclipse access to the knowledge and staggering resources of Fujitsu & Toyota. No other speaker company in the world, even the most famous ones, can call on this level of background expertise. Like race cars, speakers can be tuned and improved over time and the Eclipse design team have spent the last six years doing exactly that to deliver the latest 2014 TD models.

TD510MK2

Stereo Monitor
£1,200 (pair)

To varying degrees the sound from most loudspeakers is colored by the flapping panels and resonance of a box shaped wooden or plastic cabinet. The egg-shaped mineral-loaded resin cabinet of Eclipse speakers, provides the most rigid form known to nature and the internal construction and design inhibits resonances. All unnecessary energy is earthed through the internal zinc-alloy structure to the stand. Eclipse TD speakers are built to let you hear only the accurate signal from the drive unit, with the cabinet itself adding nothing.

Driver	3.9 inch full range fibreglass
Frequency Response	42Hz – 22kHz
Sensitivity	84dB
Nominal Impedance	6 ohms
Angle Adjustment	10 degrees – 75 degrees
Options	Stands, Brackets
Dimensions (W x H x D)	10 x 15.3 x 15 inches / 255 x 391 x 381 mm
Weight	20.9lbs / 9.5kgs
Available Finishes	Black, White, Silver
Grills	Black, White, Gray

Engelholm Audio

Sweden
www.engelholmaudio.com

For over a decade, Engelholm Audio has designed, developed and manufactured loudspeakers, electronics and room acoustics. The goal has been, and will always be, to exceed any listeners expectation. The Engelholm Audio products are set to meet the highest standards and combine new manufacturing technologies as well as exciting material combinations. At the end, key design principles like controlled dispersion, materials, phase intelligent acoustics and, the so often forgotten, time aspect are all just means to ensure the listening experience.

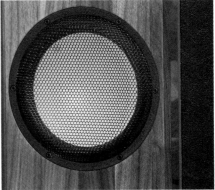

Staff

3-Way Loudspeaker
$13,300 (pair)

Based on the success of the award winning Solo, reviewers favorite Solo M and the highly rated Trill, there was still a gap in the Engelholm Audio's product range. For those with a bigger room and a musical taste that demands the low frequency to be really true to the recording, we are proud to present Engelholm Audio Staff. Staff has a lot in common with its siblings, but the 9 inch ceramic sandwich woofer truly sets it apart! The sonic impression is superb with a better transparency in the mid range and a more powerful low end.

Type	Bass-ported, Mid closed
Frequency Response	20Hz – 40kHz
Crossover	97Hz, 2.1kHz
Sensitivity	87dB
Dimensions (W x H x D)	10.2 x 39.3 x 16.1 inches / 260 x 1000 x 410 mm
Weight	114.6lbs / 52kgs
Available Finishes	Walnut, Oak with Black or White leather
Grill	Black

Eryk S Concept

Poland
www.eryksc.com

The Founder, Eryk, has spent twenty years seeking better design parameters for home audio systems, much of his research in this sphere concentrating on unique design concepts and systems configurations. He treats his designs as musical instruments, whose every string must be tuned to perfect harmony with others. With this in mind he concentrates on the object and sound transmission, the two hearts of every audio system. Eryk's audio designs have unique shapes based on good taste and ergonomics.

Superioro

Desktop Speaker
$1,490 (pair)

They call it most the innovative concept: elegant & modern in appeal, made from eco materials, nicely priced, desktop / portable with hi-end drivers and foremost full custom finished. Art graphics are created by selected group of artists. Superioro gathers art & hi-tech design in one. With Superioro you get 100% custom ideas. The company would be pleased to help you create your speaker. Superioro is a unique true art box with hi-end components. For Eryk, loud-speakers are not an end in themselves but a means to an end – the reception of music in the best possible conditions.

Music Power	30W
Nominal Impedance	4 ohms
Frequency Response	40Hz – 24kHz
Sensitivity	87dB
Dimensions (W x H x D)	4.3 x 10.6 x 10.6 inches / 110 x 270 x 270 mm
Available Finishes	Any Color. Any Design.

Everything But The Box

Bulgaria
www.ebtb.eu

Everything But The Box is a company and a brand that combines uncompromising acoustics and elegant style with supreme attention to detail. In all their new products, they further develop the handcrafting tradition and use of high-grade materials, thus defending their world reputation of a manufacturer of exclusive acoustic systems that stand out from the products in their price range. Everything But The Box loudspeakers distinguish themselves from competition by their state-of-the-art design and natural and transparent sound.

  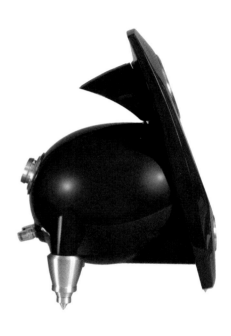

Terra Mk3
2-Way Loudspeaker
$2,599 (pair)

Terra Mk3 is a revolutionary High End Lifestyle loudspeaker. Compact dimensions, perfect proportions, and pure lines give the Terra Mk3 unmistakable presence. Terra Mk3 fuses advanced technology with staggering design to create possibly the world's most desirable art form. Based on a sophisticated Aluminum enclosure with a hand-crafted composite baffle, Terra Mk3 delivers stunning performance for all kinds of music. Amazingly deep bass, surprisingly transparent mids, and clear and non-fatiguing high frequencies ensure that Terra Mk3 has style and substance.

Drivers	30 mm Silk Neodymium tweeter, 4 inch Neodymium shielded bass-mid
Frequency Response	59Hz – 25kHz
Nominal Impedance	8 ohms
Sensitivity	86dB
Power	150W
Dimensions (W x H x D)	8.2 x 15 x 11 inches / 208 x 381 x 279 mm
Weight	11lbs / 5kgs
Available Finishes	16,000 colors

Ferguson Hill

United Kingdom
www.fergusonhill.co.uk

Founded in 2003 by former aeronautical engineer Timothy Hill, the company arrived on the hi-fi audio scene with the design, manufacture and launch of the groundbreaking FH001, the world's first ever front loaded horn dipole loudspeaker. The FH001 superb sound and eye catching design has garnered many fans and can be found in luxury homes, offices and hotels across the globe from California to Hong Kong.

FH001

Horn Speaker
£17,144 (pair)

The Ferguson Hill loudspeakers are full range units, with no crossover, however, a sub bass unit is required below 150Hz. They reproduce music with a high level of clarity, detail and dynamics, are highly efficient, and will go loud with just a few watts (5 watts). This allows their use with relatively low powered amplifiers, further increasing the level of clarity obtainable, from such relatively 'simple' amplifier circuits.

Driver	Modified Lowther DX3
Frequency Response	150Hz – 20kHz
Sensitivity	100dB
Nominal Impedance	8 ohms
Recommended Amplification	3W – 50W
Bass Options	FH002 bass unit and FH003 bass amp for £5,610
Material	Clear Acrylic
Thickness	0.31 inches / 8 mm
Dimensions (W x H x D)	36.2 x 64.9 x 28.3 inches / 920 x 1649 x 719 mm
Weight	59.5lbs / 27kgs
Available Finishes	Transparent

Gato Audio

Denmark
www.gato-audio.com

Designed and built with beauty, simplicity, functionality and an extremely elegant user interface, Gato Audio electronics possess both brute strength and finesse to bring the subtlest musical details to life with precision and authority. Gato Audio loudspeakers are lavishly constructed of the finest materials with laminated cabinets, state of the art Danish high end drivers, and carefully calibrated heavy-duty crossover networks to provide breath-taking concert hall performance in a beautiful yet compact form.

FM-6

2.5-Way Loudspeaker
$14,000 (pair)

The 2.5-way floorstanding speaker offers effortless musical reproduction in medium to large listening rooms. The FM-6 incorporates several technologies of Gato Audio's proprietary Resonance Eliminating Program. Top-class crossover such as huge low-resistance insatiable inductor coils, costly Clarity Cap and induction-free power resistors are mounted on fiber-glass PCB and then casted into resonance dampening resin. The best of Danish drivers have been selected, modified and adopted in the final design.

Drivers	30 mm Ring Radiator (HF), 2 x 180 mm Sliced Cone (MF & LF)
Frequency Response	38Hz - 30kHz
Sensitivity	90dB
Recommended Amplification	50W - 300W
Nominal Impedance	4 ohms
Connectors	Gold plated WBT bonding post / 4 mm plugs, bi-wire
Dimensions (W x H x D)	13 x 40 x 14.5 inches / 300 x 1020 x 370 mm
Weight	66.1lbs / 30kgs
Available Finishes	High Gloss - Black, Walnut, White

Genesis Advanced Tech.

Washington, USA
www.genesisloudspeakers.com

Genesis Advanced Technologies is a leading developer of high-end hi-fi stereo systems. From cost-no-object high-end loudspeakers to high performance, high efficiency power amplifiers to Absolute Fidelity® component, loudspeaker, and power interface cables, Genesis shows its commitment to the absolute best. Genesis products show uncompromising quality throughout the range utilizing the same driver and crossover components in every product from the exclusive, custom flagship loudspeaker system down to the affordable, petite stand-mounted loudspeaker.

Genesis 1 Dragon

Loudspeaker
$350,000

The Genesis 1 "Dragon" is the latest iteration of the flagship loudspeaker system from Genesis. Custom built in the USA for the music lover who accepts no compromise in the reproduction of the musical event, this state-of-the-art system is designed to reproduce music at live listening levels with virtually no restrictions on dynamic range, frequency response, or width, breadth and height of soundstage. The four tower Genesis 1 loudspeaker system is one of the very few speakers in the world that is capable of recreating the size, weight, impact and listening levels of a live performance with ease.

Drivers	26 x 1 inch tweeters, 12 x 12 inch woofers, 75 inch ribbon midrange
Frequency Response	16Hz - 40kHz
Sensitivity	91dB
Nominal Impedance	4 ohms
Inputs	XLR
Power Rating	400W per 6 channels
Dimensions (W x H x D)	Mid/Tweeter Panel: 41.5 x 90 x 5 inches / 1054 x 2286 x 127 mm
	Woofer Tower: 15.5 x 90 x 19 inches / 394 x 2286 x 483 mm
Weight	992lbs / 450kgs
Available Finishes	Black, Rosewood, Corian

German Physiks

Germany
www.german-physiks.com

All German Physiks loudspeakers use a unique DDD Driver, which can run from 40Hz to 24kHz, so there is no nasty crossover in the crucial mid-range region, where human hearing is most sensitive. This enables their loudspeakers to provide palpably realistic stereo images, with new standards of speed, transparency and above all musicality. Unlike conventional designs, the stereo images their loudspeakers produce can be enjoyed from a wide range of listening positions, making them very family friendly.

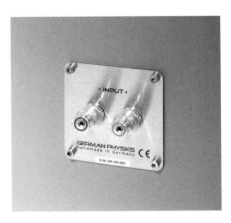

Unlimited Mk II

2-Way Loudspeaker
$13,500 (pair)

The Unlimited Mk II is our entry-level model and has received numerous excellent reviews worldwide. It uses a single carbon fibre DDD driver mounted on top of a slim floor standing cabinet, with a downward firing woofer set in its base. The Unlimited MK II provides the famous German Physiks trademark dynamics, transparency and musicality, but at a new lower price point. A footprint of only 9.5 inches square make it very easy to position and set-up. It is exceptional value for money and makes the unique enjoyment that German Physiks loudspeakers can provide available to a much broader audience.

Drivers	1 x Carbon fibre DDD, 1 x 8 inch woofer
Frequency Response	32Hz – 24kHz
Nominal Impedance	4 ohms
Recommended Amplification	90W
Sensitivity	88dB
Power Handling	110W – 170W
Crossover Frequency	200Hz
Input Connections	1 set of binding posts
Dimensions (W x H x D)	9.5 x 41.3 x 9.5 inches / 240 x 1050 x 240 mm
Weight	63.7lbs / 28.9kgs
Available Finishes	Satin finish: White, Black, Light Grey, Dark Brown

GoldenEar Technology

Maryland, USA
www.goldenear.com

In this industry, a "golden ear" is someone who hears exceptionally well. The GoldenEar team fully utilizes all its combined experience to deliver more sonic superiority, with performance that is often unequalled in competitors' speakers selling for three, four or five times as much as a comparable GoldenEar speaker. GoldenEar does it because they can – and because they know that you will enjoy and appreciate the result.

Triton Seven
Tower Speaker
$699.99 (pair)

Although it is the first Triton Tower without a built-in powered subwoofer, it still achieves superbly deep, subwoofer-like impactful bass performance by combining advanced technology drivers with their highly evolved and sophisticated bass-loading techniques. The Seven's unique cabinet shape is both acoustically purposeful as well as strikingly attractive.

Drivers	Mid/Bass: 5.25 inch High-Definition cast
	8 inch Planar sub-bass radiators
	Tweeter: HVFR™ High-Velocity Folded
Frequency Response	29Hz - 35kHz
Efficiency	89dB
Nominal Impedance	8 ohms
Recommended Amplification	10W - 300W per channel
Base Dimensions (W x D)	10.5 x 14.5 inches / 266 x 368 mm
Dimensions (W x H x D)	5.75 x 39.75 x 11 inches / 146 x 1010 x 279 mm
Weight	32lbs / 14.5kgs
Available Finishes	Black

Gradient

Finland
www.gradient.fi

Gradient has been designing and manufacturing loudspeakers since 1984. From the founding date of Gradient, their aim has been to minimize the room-loudspeaker interaction. This has led to designs, totally unique and new in the era of loudspeaker manufacturing. Gradient is a trademark of Gradient Labs Ltd, Finnish family-owned company. The headquarters of the company are located in Porvoo, a beautiful old town having hundreds of years of heritage of craftsmanship.

Helsinki 1.5
3-Way Loudspeaker
$6,495 (pair)

Gradient Helsinki is tailored to produce the most beautiful sound in all environments from home to lounge bars and hotel lobbies. These loudspeakers are easy to position and they deliver very open and wide sound stage everywhere to the room. Gradient Helsinki utilizes Controlled Directivity, the very essence of unique Gradient loudspeaker design philosophy. By directing more of the speakers high frequencies and middle frequencies to the listener and less to the reflecting surfaces the listener hears more of the original recording and less of the room.

Drivers	300mm woofer, 150mm midrange, 19mm tweeter
Frequency Response	60Hz – 20kHz
Nominal Impedance	6 ohms
Sensitivity	85dB
Recommended Amplification	50W – 250W
Crossover	200Hz; 2.2kHz
Features	Bi-wiring, bi-amping ready
Radiation Patterns	Bass: Dipole; Midrange & Treble: Cardioid
Dimensions (W x H xD)	13.7 x 36.2 x 19.6 inches / 348 x 919 x 498 mm
Weight	50.7lbs / 23kgs
Available Finishes	Black, White, Gray, Bamboo, Birch, Cherry Veneer, Oak, Walnut

Grado

New York, USA
www.gradolabs.com

Grado, one of the oldest family owned companies in the Audio Industry, has for over half a century been the leaders in design engineering for the high-end audio and recording industries. Grado is famous for their remarkable headphone and phono cartridge designs and holds over 48 patents. Time honored manufacturing know how and painstaking attention to design detail allows previously unobtained levels of pitch control, harmonic accuracy and bass quality to be achieved.

PS500

On-Ear Headphones
$595

There is art in the heart and soul of every Grado product. The PS500 truly conveys the heart and soul of every musical performance you will listen to through them. This is the result of the art learned over the course of almost 60 years of designing and manufacturing products that faithfully follows music. A compact monitoring tool that will put a smile on the face of the most demanding music professionals as well of demanding audiophiles.

Features	Vented Diaphragm, Hybrid air chamber, UHPLC copper voice coil wire, UHPLC copper connecting cord
Transducer Type	Dynamic
Operating Principle	Open air
Frequency Response	14Hz - 29kHz
Sensitivity	98dB
Nominal Impedance	32 ohms
Driver matched	0.05dB

HHR Exotics

Ohio, USA
www.hhr-exoticspeakers.com

HHR Exotic Speakers is a small division of H & H Research, an established Gas laser company started some 30 years ago. While HHR Exotics is just two years old as a commercial enterprise, they began from strong roots in the laser and audio industries. Their sole motivation is to provide the highest quality components and workmanship with the greatest return for your investment available today. As owner of HHR Exotics, it is Dale Harder's desire to provide audio enthusiasts the finest loudspeakers ever developed.

Walsh TLS-I
Loudspeaker
$12,000 (pair)

More than 30 years of advancements in technology and materials allows HHR Exotics to produce a Walsh style driver of unsurpassed sound quality and exotic beauty. The TLS-1 is a vastly improved version of the original Ohm Walsh "F". While the design of the TLS-1 is modeled after the old "F" series, it currently incorporates more than 50 design improvements that are the direct result of 35 years of research and experience. The New TLS- 1 is intended for use in medium to large rooms from 200 to 400 square feet.

Driver	12 inch diameter, 2.1lbs Alnico, V-7 Slug, 10.25lbs magnet, 10,000 gauss
Frequency Response	29Hz – 22kHz
Nominal Impedance	6 ohms
Sensitivity	89dB
System Resonance	24.5Hz
Maximum Power	150W
Protective Fuse	AGC4
Dimensions (W x H x D)	19.2 x 44 x 19.2 inches / 489 x 1118 x 489 mm
Weight	150lbs / 68kgs
Available Finishes	Hardwood veneers finished in satin or high gloss tung oil
Grill	Black

Horning Hybrid

Denmark
www.horninghybrid.com

Amongst the fine arts, music seems unique in its ability to stir our emotions, whether at live performances, from media broadcasts or in recordings that we all can enjoy at home. The company is particularly proud that Horning has achieved their status amongst knowledgeable enthusiasts without hype, advertising or marketing, they convince on hearing. The performance of their products is such that its demands are just as great; only the finest sources and amplifiers will do it justice.

Euphrodite
Loudspeaker
$22,000 (pair)

For two decades Horning loudspeakers have successfully evolved to exceed the ever increasing expectations of the most discerning music lovers. The Horning Eufrodite full range loudspeaker is the pinnacle of that objective. The Eufrodite is both sophisticated yet surprisingly simple, it is the manifestation of a deep appreciation of music translated through an ambitious approach to acoustics into a desirably sensitive and efficient loudspeaker. The notion of Hi Fi becomes superfluous. When confronted with your favorite musical moments with heretofore only imagined realism, your body will take over. Your foot will begin to tap and you'll be lost in the musical moment.

Drivers	6dB pr octave tweeter, 6dB-12dB pr octave mid, 6dBSA >> pr octave bass
Frequency Response	28Hz - 20kHz
Nominal Impedance	8 ohms
Power	30W
Efficiency	98dB
Wiring	Horning Reference Cable
Dimensions (W x H x D)	8.6 x 47.2 x 25.7 inches / 220 x 1200 x 650 mm
Weight	143lbs / 65kgs

Kaiser Acoustics

Germany
www.kaiser-acoustics.com

Kaiser Acoustics GmbH was founded four generations ago and remains a family owned and operated business today. Kaiser Acoustics has an established track record in the field of acoustic installation for some of the world's most prestigious public buildings, private yachts, sound recording studios, and conference rooms. There is undoubtedly a huge sense of pride within this company, based on the classic Bavarian love of beautiful materials and exquisite craftsmanship.

Kawero! Vivace

3-Way Loudspeaker
$56,000 (pair)

This loudspeaker sounds rich and transparent, with a body of sound that seems near impossible given its relatively compact proportions. Kawero! comes custom-tailored to your needs. You can select from a great variety of veneers. Acoustic fine-tuning is individual and there are three custom woofers to choose from. You can also consider the external crossover option which brings further significant improvements to the whole speaker package. The Kawero! brings a marriage of unequalled workmanship and performance together.

Design	3-way (reflex loaded midrange)
Bandwidth	29Hz - 26kHz
Sensitivity	88dB
Minimum Impedance	4 ohms
Nominal Impedance	6 ohms
Recommended Power	50W
Connectors	Single wiring (bi-wiring on request)
Dimensions (W x H x D)	13 x 47.6 x 19.4 inches / 330 x 1210 x 493 mm
Weight	218.3lbs / 82kgs
Available Finishes	Any. Special and High Gloss lacquerings available.
Distributor in US	www.lessloss.com

Kharma

Netherlands
www.kharma.com

Kharma embodies the seeking to ultimate beauty, where Kharma converges audiophile excellence with aesthetic beauty. A genuine approach to reveal the most subtle musical experience and to unlock immeasurable aesthetic joyful inner experiences in the listener. Like a piece of art in painting can evoke feelings of beauty and excitement in the viewer, the Kharma products are pieces of art in audio vibrations and they will evoke the highest feelings of beauty and wonder in the listener.

Elegance dB9

3-way Loudspeaker
$36,000 (pair)

The Elegance collection is succeeding the very successful Ceramique line of high-end loudspeakers from Kharma. The Ceramique loudspeaker line was introduced in 1998 and the use of Ceramique-based drivers was the reason for naming this series like they did. The Elegance dB9 was launched with the Kharma Composite driver replacing the Ceramique drivers from the Ceramique line. However, as eager as Kharma is to strive to perfection in everything we do, Kharma already has developed a signature version of the dB9, called the dB9-S. This new dB9-S has, amongst others, a new Omega 7 driver for the midrange and extra cross section for the silver internal wiring.

Drivers	True Beryllium tweeter, Kharma Composite Driver, Aluminum woofer
Type	EL-dB9-1.0
RMS Power	250W
Program Power	500W
Frequency Response	26Hz - 30kHz
Nominal Impedance	4 ohms
Sensitivity	89dB
Dimensions (WxHxD)	12.5 x 38.6 x 23.8 inches / 317 x 981 x 606 mm
Weight	168lbs / 76.2kgs
Available Finishes	14 color options available including Black, White, Brown

Kralk Audio

England
www.kralkaudio.com

Kralk Audio is a small independent loudspeaker manufacturer based in Stanley, West Yorkshire. Their aim is to give customers the highest quality audio experience utilizing the best components, and build quality possible. Their loudspeakers are hand built to exacting tolerances, and are assembled with care and meticulous attention to detail, so as to afford each and every customer a thrilling and satisfying aural experience throughout their range of products.

DTLPS1

Bookshelf Speaker
$1,173 (pair)

The DTLPS1 is a compact 2-way bookshelf / stand mounted speaker featuring the Advanced Transmission Line Port System for outstanding low frequency response at low and high volumes. The 165mm bass driver is housed in a most rigid cabinet for almost zero coloration of the system for superb controlled low bass response you can feel as well as hear, the doped paper cone and vented coil pole help give the system very clear and detailed midrange performance, exceptional stereo image performance and life-like dynamics that are not normal with a speaker of this size.

Crossover Frequency	2.5kHz
Frequency Response	35Hz - 22kHz
Nominal Impedance	8 ohms
Input Connectors	2 pairs 4 mm sockets (bi-wire or bi-amp)
Recommended Amplification	20W - 150W
Sensitivity	89dB
Internal Volume	15.3 liters
Optional	Extra Bass Extender Stand XB-1 for $546
Dimensions (WxHxD)	7.2 x 11.6 x 11 inches/ 185 x 296 x 280 mm
Weight	13.7lbs / 6.2kgs
Available Finishes	Walnut, Sapelle, Black Ash, Light Oak

Legacy Audio

Illinois, USA
www.legacyaudio.com

Legacy Audio handcrafts the finest loudspeakers available. Their speakers are renowned for establishing reference level performance in home theater and audiophile settings and their 7 year performance guarantee assures your Legacy speakers will provide the highest level of enjoyment. A product platform unequalled in terms of aesthetics, build quality, sonic performance and technology. An uncanny understanding of physics and fluid modeling, combined with good old-fashioned hard work and craftsmanship.

Aeris

4.5-Way Loudspeaker
$15,900 (pair)

Aeris from Legacy Audio is a speaker system whose striking looks are matched only by its performance capabilities. This 4.5-way system is the first to use the Legacy Dual Air Motion Tweeter (an advanced AMT design) and features open air dipole midrange, dual subwoofers with internal bass amplification, room correction and craftsmanship display options. A natural fullness in the treble is exhibited, reminding us why we love high-end audio so much.

Drivers	6 drivers
Frequency Response	18Hz - 30kHz
Nominal Impedance	4 ohms
Sensitivity	95.4dB
Binding posts	2 pairs
Crossover Frequency	80Hz, 2.8kHz, 8kHz
Recommended Amplification	60W - 500W
Dimensions (W x H x D)	Upper Cabinet: 14.5 x 58 x 16 inches / 368 x 1473 x 406 mm
	Base: 19 x 1 x 17.5 inches / 483 x 25 x 445 mm
Weight	171lbs / 77.6kgs
Available Finishes	11 choices

Mansberg

Holland
www.mansbergsound.com

Mansberg is a Dutch manufacturer and distributor of exclusive state-of-the-art audiophile speakers. This young and dynamic company was started in 2009 by Marius Dielemans and Rene Bovenberg who share a common background in industrial design, marketing and hifi. They believe there is more than enough-robot-assembled, computer-controlled, imagination-free plastic throw-away stuff pretending to be audio gear. Mansberg has decided differently, on all counts. Ruthlessly.

Sinus

4-Way Loudspeaker
$18,500 (pair)

Top of the line is the Sinus speaker which is a 4-way system containing five drivers and offers a compelling combination of striking design and remarkable sound quality. They are listed in the Audio Hall of Fame for good reason; Passion inspires, provokes and invigorates! The Sinus is designed in such a way that the shape does not compromise the sound but in fact enhances it. The Mansberg Sinus performs with enough strength to reveal true nature of guitar plucks, organ notes and violin strings. This is where listening translates into pure pleasure.

Drivers	5 drivers
Frequency Response	22Hz – 20kHz
Sensitivity	91dB
Nominal Impedance	4 ohms
Crossover	80Hz, 750Hz, 4.5kHz
Power Handling	150W/250W
Dimensions (W x H x D)	10.2 x 60.6 x 18.5 inches / 260 x 1540 x 470 mm
Weight	99.2lbs / 45kgs
Available Finishes	Piano Black, Grey, Blue, Red Saraya, Oak Veneer and other colors

Mark & Daniel

Arizona, USA
www.mark-daniel.com

Mark & Daniel Audio Labs has designed and manufactured their high-end audio products since 2004, specializing in R&D for sound drivers and speaker systems as well as the CAM solid surfacing applications with the goal of creating systems capable of true hi-fi reproduction. M&D has successfully overcome three primary restrictions in the development of hi-fi speaker systems today - Narrow operating bandwidth for tweeters; Insufficient driving dynamics to woofers; Lack of FMD control. M&D has pioneered a "Conceptional Revolution" in high fidelity music reproduction.

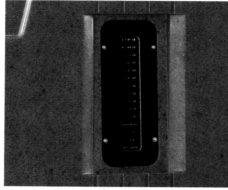

Fantasia-Cello

Floor-standing Loudspeaker
$22,500 (pair)

The Fantasia-Cello sound system is the most advanced loudspeaker designed and engineered by Mark & Daniel, employing the best of their pioneering concepts and state-of-the-art technologies. While most loudspeaker system manufacturers resort to dual woofers as the only viable measure to achieve low frequency augmentation and sound pressure, yet not enough to reach true deep bass extension, resulting in the necessity of sub-woofers; Mark & Daniel daringly push the envelope of their super woofer design and thereby persistently maintain minimal phase distortion through one single woofer. It is a truly exceptional integrated speaker system optimizing musicality, high definition and effortless output capability.

Frequency Response	20Hz - 35kHz
Crossover Frequency	650Hz
Nominal Impedance	4 ohms
Sensitivity	87dB
Power Handling	200W
Enclosure Material	C.A. Marble Solid Surface
Dimensions (WxHxD)	16.5 x 42.1 x 11 inches / 419 x 1069 x 279 mm
Weight	139lbs / 63kgs
Available Finishes	Majestic Granite and Starry Black
Grill	Black

Morel

Israel
www.morelhifi.com

Morel, an international leader in speaker components and systems since 1975, manufactures handcrafted, award-winning speakers and audio drivers for the mid to high-end OEM, home and car audio markets. High standard for technological innovation & design excellence, makes Morel audio speakers the choice of many in the music industry. At the heat of the Sopran are new 6-inch titanium based drivers, that seamlessly integrate with a sophisticated crossover architecture to preserve the drama and dynamics of the original sound, creating a tactile musical experience.

Sopran

3-Way Loudspeaker
$12,000 (pair)

The Sopran 3-way (five drive units) is a combination of creative vision and acoustic physics. Featuring Morel's most advanced technologies; internal standing waves diffusion shaped the curved carbon fibre composite cabinet enabling no internal absorption material, creating a speaker so transparent and natural that it sounds as if there was no cabinet at all. For tight and controlled bass reproduction a new in-line port technology was developed molding the speaker's back spine appearance for a significantly improved soundstage and realism. Completing the cabinet design are the new innovative Lotus grilles.

Drivers	3 x 6-inch (160mm) new titanium former drivers, 1 x 6-inch (160mm) new titanium former driver, 1.1-inch (28 mm) Acuflex™ soft dome
Frequency Response	30Hz - 22kHz
Nominal Impedance	4 ohms
Nominal Power	250W
Peak Power Handling	1000W
Dispersion	Within 1.5dB
Sensitivity	91.5dB
Crossover Frequency	250Hz, 2.2kHz
Dimensions (W x H x D)	10 x 44.6 x 16.9 inches / 255 x 1135 x 430 mm
Weight	69.4lbs / 31.5kgs

MrSpeakers

California, USA
www.mrspeakers.com

MrSpeakers was founded with a simple goal in mind: great headphone products at affordable prices. At MrSpeakers, they believe audiophile performance should not require stratospheric prices, and as such they strive to deliver incredible performance and high value for each and every product they deliver. MrSpeakers was founded by Dan Clark, an electrical engineer who has been working in and around the high-end audio market for more than 20 years. At MrSpeakers, every headphone is hand assembled and tuned to ensure the highest possible performance.

Alpha Dog

Over-Ear Headphone
$599.99

MrSpeakers flagship product is the Alpha Dog. The Alpha Dog revolutionizes close-back headphones by using 3D printing to create a double-walled enclosure that simultaneously improves sound quality and increases isolation. The Alpha Dog headphone is truly closed, yet delivers a soundstage with size and clarity that competes with the finest open backed headphones. With the Alpha Dog, you'll never again have to trade isolation for sound quality. If you are interested in reference-quality headphones with state of the art soundstage, detail retrieval, ambience, and linearity it's all about the Alpha Dog.

Frequency Response	15Hz – 18kHz
Power Handling	1500mW
Sensitivity	92dB
Options	¼ inch Single-Ended Cable, Balanced 4-Pin XLR Cable
Weight	4.5lbs / 2kgs

Musique Concrete

France
www.musique-concrete.com

During the early 2000s, Marc Henry spend several years researching about the influence of wall material, profile, and truncature of horns. His work become famous, and his brand Musique-Concrète sells horns in many countries. The very first of them were made of concrete. Since 2007, thanks to the new partners, the company goal is to ally extreme high-end audio with a unique French touch in design, careful handcrafting, beautiful materials and finishes, for a life-time satisfaction.

La Grande Castine
3-Way Horn System
$100,000

La Grande Castine is the result of a long development by the horn system maker Marc Henry and the violin soloist Hugues Borsarello. As any other music lover, Hugues loves the incredible liveliness a horn system can reach. As a musician, he gives a very particular importance to coherency. More than an assembly of very performing parts, La Grande Castine is a synergy: This 3-way horn system play as a single full range speaker. Each horn and driver is vibration-free fixed, bandwidth and efficiency are mechanically adjusted to keep the crossover very simple. Technical features are as the design is: a very hard-to-get simplicity.

Type	3-way horn system + open baffle subwoofer
Drivers	Treble: 1 inch compression driver, neodymium magnet
	Midrange: 2 inch compression driver, neodymium magnet
	Bass: 15 inch neodymium magnet
Sensitivity	108dB
Frequency Response	20Hz - 20kHz
Recommended Amplification	1W - 30W
Input	Single pair of platinum binding post
Dimensions (W x H x D)	36.2 x 78.7 x 49.2 inches / 920 x 2000 x 1250 mm
Weight	287lbs / 130kg
Available Finishes	Many

Mythos Audio

Greece
www.mythosaudio.com

The History of Mythos Audio goes back to the 80's. The foundation of Mythos Audio came in 1998 and since then Mythos Audio evolved into one of the biggest loudspeaker and audio furniture manufacturer of Greece. Their love for music and the permanent search of "clean" and "crystal" sound led them to experimentations which then led them to the creation of their first loudspeakers. In their completely organized and equipped facilities each speaker and furniture is manufactured, assembled and dyed by experienced personnel and passes from technical and qualitative control.

Olon

Loudspeaker
$52,880

The name is inspired from the Ancient Greek word where Olon (the whole) was the hidden idea behind the then known, or unknown universe. With all the mysteries involved in a sophisticated elaborate cabinet construction, and a whole lot of love for music, the Olon loudspeaker was born. Olon uses, plywood stacked layers for its cabinet construction. Mythos Audio is confident enough that the Olon loudspeaker in a high-end system will travel you in the magnificent, unknown, mysterious universe that music hides within.

Drivers	12 inch Visaton, 6.5 inch Scanspeak, Mundorf AMT
Frequency Response	18Hz – 40kHz
Recommended Amplification	>50W
Sensitivity	88dB
Nominal Impedance	8 ohms
Options	Spikes, Silver Crossover, Silver caps, Silver Internal Wiring, Coils
Dimensions (W x H x D)	19.6 x 62.9 x 28.3 inches / 498 x 1598 x 719 mm
Weight	264lbs / 120kgs
Available Finishes	Plywood

Origin Live

United Kingdom
www.originlive.com

Origin Live is a highly innovative UK based company that designs and manufactures Turntables, Tonearms, Loudspeakers and cables. Founded in 1986, it has built up a wealth of experience leading to numerous awards. With it's renowned musical and entertaining sound quality – reviewers often refer to their products as "addictive" , "the type of sonics that make you keep listening to record after record". Origin Live has won group comparison tests with their speakers and have recently developed high-end hanging designs.

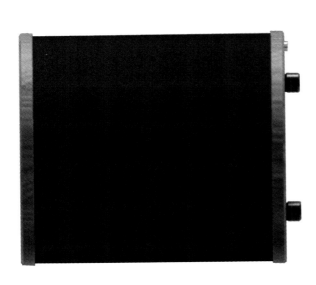

Astute 8C Ultra

Loudspeaker
$18,500 (pair)

The innovative design of the Astute Loudspeaker extends much deeper than its attractive aesthetics. This is a revolutionary design with explosive dynamics due to very high sensitivity. The co-axial drivers mounted in each cylindrical cabinet produce an amazingly natural, out of the box sound. There is a total absence of resonance due to hanging the cabinet which de-couples it from floor energy reflections present in conventional speakers. Low frequencies integrate seamlessly using a floor standing sub-woofer.

Drivers	8 inch midrange, 1 inch tweeter
Frequency Response	100Hz – 21kHz
Crossover	1.8kHz
Sensitivity	95dB
Nominal Impedance	8 ohms
Power Handling	250W
Options	Proprietary Origin Live wall brackets, ceiling brackets, floor stands
Dimensions (Diameter x Depth)	12.9 x 12.9 inches / 300 x 300 mm
Available Baffle Finishes	Havana / Neopolitan Strand Bamboo, Amber Edge Grain, Gloss Black
Available Body Finishes	Different shades of leather.

PerfectSound

Taiwan
www.perfect-sound.com.tw

Butterfly is the first impression for PerfectSound. There are three curves of butterfly in both left and right, representing the low, midrange, and high frequencies in the stereos that are linked and cannot be missing. Just like PerfectSound insistence to hold each range characteristic and present most real affected integrity for you. PerfectSound would like to lead you to roam through the music domain like butterfly flutters into the note.

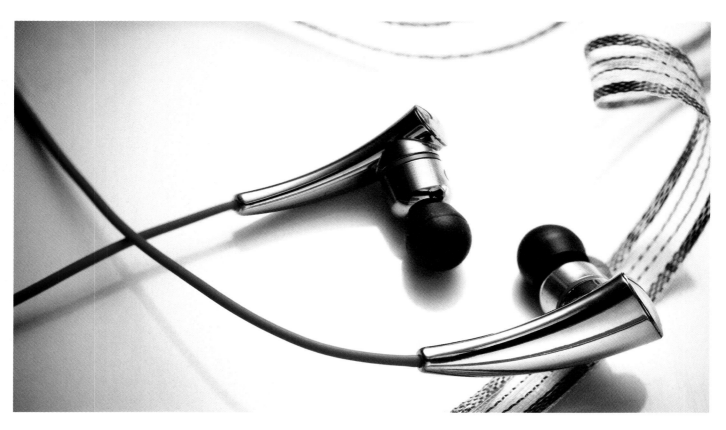

S102

In-Ear Headphone
$200

PerfectSound would like to introduce to you their sensual S series range. PerfectSound's S102 has a unique appearance and aluminum-magnesium alloy cavity. It has been molded in the shape of a horn representing the perfect uniqueness of design in a world of similarities. Apart from its trend setting design the S102 makes crystal clear sounds. Precision crafted from Taiwan, it brings you elegant and high sound quality. The perfect sound from PerfectSound. You will just love these earphones. Stand out from the crowd, be unique. Your style. Your music.

Driver	Dynamic
Impedance	18 ohms
Connector Type	3.5 mm
Cord Length	1.2 meter
Options	3 sizes of ear plugs, leather storage bag
Driver Diameter	0.3 inches / 8mm
Weight	0.04lbs / 19gms

PerfectSound

Taiwan
www.perfect-sound.com.tw

Butterfly is the first impression for PerfectSound. There are three curves of butterfly in both left and right, representing the low, midrange, and high frequencies in the stereos that are linked and cannot be missing. Just like PerfectSound insistence to hold each range characteristic and present most real affected integrity for you. PerfectSound would like to lead you to roam through the music domain like butterfly flutters into the note.

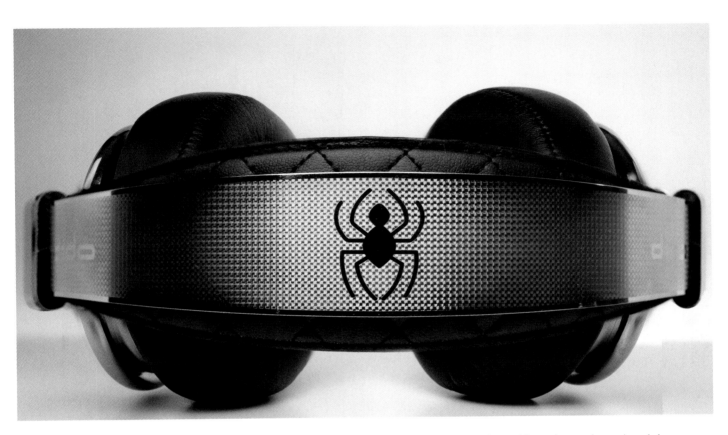

d901

Over-Ear Headphone
$780

The D series is Perfect Sound's reference range. Everything about these headphones are exquisite, from the luxurious design through to the beautiful sound reproduction, a sound that will relax you and brighten up your day. The d901 comes with a metal headphone rack, lightweight and durable material, silver exterior. A protein leather headband, comfortable earmuffs and foldable ear cups are its other features. It has 2 input cable ports. So you can plug your headphone cable in to the right or left ear. Surround yourself with dido. It is designed to make you feel comfortable and elegant when listening to music.

Driver	Dynamic
Impedance	16 ohms
Connector type	3.5mm
Cord Length	1.3 meter
Driver Diameter	1.5 inches / 40mm
Weight	0.9lbs / 395g
Available Finishes	Black, White

Sanders Sound Systems

Colorado, USA
www.sanderssoundsystems.com

Sanders Sound Systems has been designing Electrostatic Loudspeakers and the amplifiers to drive them since 1974. Their reputation for honesty and dedication to customer service is due to their belief that satisfying customers is the best way to build a successful company. They offer a 30-day, in-home, risk-free trial. They will ship you any equipment that you wish. You can use it in your own home to listen to your familiar music, in your own listening room, with your own associated audio components for up to 30 days.

Model 10

Electrostatic Loudpeaker
$14,000* (pair)

This Sanders' flag ship speaker. It is the best speaker Roger Sanders knows how to make after 30+ years of research, design and testing of electrostatic speakers. This is simply a no-compromise electrostatic speaker that solves all the problems that plague other electrostatic speakers. This speaker practically disappears into the room, both sonically and visually. Roger's "Ultrastat" panels will play at ear-bleeding levels, cannot be arced, do not use or need protective circuitry, are immune to humidity, dust, and dirt, and do not need or use dust covers.
Price includes Electrostatic Speakers, Magtech Amplifier, Digital Crossover.

Driver	Bass: 10 inch Transmission line
Frequency Response	20Hz – 27kHz
Crossover Frequency	172kHz
Sensitivity	94dB
ESL Power Handling	Unlimited for any amplifier intended for domestic use
Harmonic Distortion	<0.004%
Signal-to-Noise Ratio	>96dB
ESL panel size (W x H)	15 x 42 inches / 381 x 1067 mm
Dimensions (W x H x D)	15 x 69 x 18 inches / 381 x 1753 x 457 mm
Weight	80lbs / 36kgs
Available Finishes	Black, Natural Cherry, Natural Walnut

Shape Audio

Sweden
www.shapeaudio.com

The story of Shape Audio began in 2007 when friends and business partners Luciano Pasquariello and Bernt Böhmer together established a need to reinvent the conservative audio industry. Both passionate about high quality sound as well as aesthetics, they set out on a quest to find a solution to their common goal: to provide the market with exquisite design as well as an unparalleled listening experience. Ranging from its signature shape, its exclusive materials and its hand-selected technology, all factors play an equally important role in creating the ultimate sound experience.

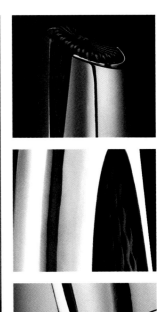

Organic Harmony
Omni-Directional Loudspeaker
$105,000

Shape Audio goes above and beyond anything the audio world has ever seen. Each beautifully finished sculpture is cast by hand in exquisite metals such as bronze, silver and gold. The Organic Harmony's unique composition enables perfect sound from a single unit. Each Organic Harmony contains an advanced and integrated audio system. Each Organic Harmony is numbered and manufactured in an exclusive limited edition of 99 units in bronze, 5 in silver and 1 in solid gold. The Organic Harmony music sculpture is a complete music system with built-in amplifiers, USB and Ethernet connection, as well as room correction and much more.

Inputs / Outputs	RCA, 3.5mm jack, S/P DIF, Toslink, USB, Ethernet, XLR
Power	230VAC; 120VAC
Frequency Response	40Hz – 30kHz
Amplification	Class-D
Efficiency	95%
Dimensions (W x H)	10.6 x 53 inches / 270 x 1350 mm
Weight	209lbs / 95kgs; 253lbs / 115kgs; 473lbs / 215kgs
Available Finishes	Bronze, Argentium Sterling Silver, 18 karat Gold ($13,000,000)

Shure

Illinois, USA
www.shure.com

With a history of audio innovation spanning over 80 years, Shure has turned a passion for making great microphones and audio electronics into an obsession. No wonder Shure continues to set the worldwide industry standard for superior microphones and audio electronics. Engineered to the same exacting standards that have made Shure the choice of pros, Shure Sound Isolating Earphones deliver unmatched performance - whether you're on stage, or simply on the go.

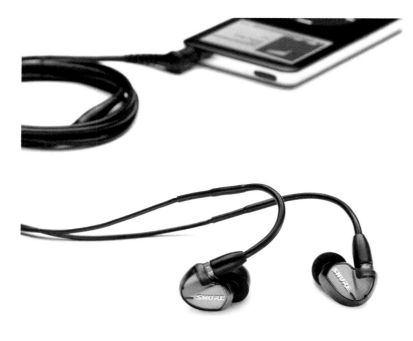

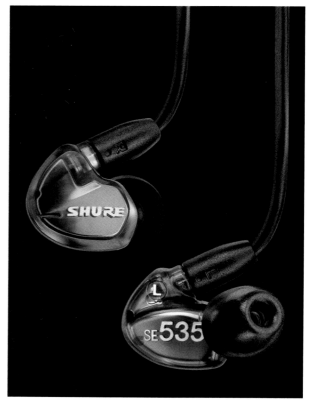

SE535

Sound Isolating Earphone
$499.99

The choice for discriminating professionals and audiophiles, the SE535 utilizes dedicated tweeter and dual woofers, for incredibly spacious sound with rich bass. The role of the included sound isolating sleeves is two-fold: blocking ambient noise and ensuring a comfortable, customized fit. A good seal is key to optimizing sound isolation and bass response as well as maximizing comfort during extended wear. Evolved from personal monitor technology, road-tested by pro musicians and fine-tuned by Shure engineers, SE535 earphones deliver an unparalleled listening experience allowing you to hear the details of your music like never before.

Driver	Triple High-Definition MicroDrivers
Frequency Response	18Hz - 19kHz
Sensitivity	119dB
Nominal Impedance	36 ohms
Detachable Cable	64 inch / 1626 mm (with wireform fit)
Inclusions	Sleeves, Carrying case, Adapters, Volume Control
Sleeve sizes	Small, Medium, Large (S, M, L)
Available Finishes	Metallic Bronze, Clear

Sonus Faber

Italy
www.sonusfaber.com

Sonus Faber for 30 years has created the perfect combination of cutting edge technology, skilled craftsmanship and elegant design. It is able to ride the key trends in the world of production and distribution of high-end acoustic speakers, and to consolidate the role of its brand as a leader in the global market. Sonus Faber has a range of approximately 30 speaker models. The incredible clarity means that they are both desirable and objects of interest for the world high-end audio community and also for music lovers, who considers them to be true musical instruments.

Olympica II

3-Way Loudspeaker
$10,000 (pair)

The Sonus Faber Olympica project is made up of a new family of passive acoustic speakers that will add to the existing catalogue from the Vicenza factory, characterized by the use of two iconic Sonus Faber materials: walnut wood and leather. With this new project, Sonus Faber has been able to balance quintessential style and significant elements of technological innovation. The models of the Sonus Faber Olympica line are entirely new. No component, element or design has been borrowed from previous models. Rather, cumulative experience and technology have been used to build innovative new cabinet structures and high performance drivers and crossover.

Drivers	Tweeter: "Arrow Point" DAD 29XTR2
	Midrange: MW15XTR
	Woofers: W18XTR
Frequency Response	40Hz - 30kHz
Sensitivity	88 dB
Nominal Impedance	4 ohms
Recommended Amplification	50W – 250 W
Maximum Input Voltage	20Vrms
Dimensions (W x H x D)	14.5 x 41.5 x 18.5 inches / 370 x 1055 x 472 mm
Weight	74.9lbs / 34kgs
Available Finishes	Natural Walnut, Graphite with leather

Sound& Design

Italy
www.sound-design.it

Sound&Design products are more comparable to musical instruments by lute-making than hi-tech stuff and they are built with the same approach. Over 90% of the production times is attributable to manual activities not automatable. The result is a natural sound that has the appearance of an instrument. Sound&Design uses the most modern technology as well, developing sophisticated software tools to simulate out cabinets and high precision CNCs for semi-finished parts that will be expertly assembled and finished by their cabinetmakers.

Tethys

3-Way Loudspeaker
$13,000 (pair)

Sound&Design's continuous quest for excellence has led them to define a new standard among floor loudspeakers. Tethys required more than 4 years long research and fine-tuning. Tethys is characterized technically by a range of solutions starting from the narrow speakers and filter components selection up to the choice of series crossover filter and the design of a waveguide for the rear emission in order to obtain a perfect energy room response. With 90 db@1m sensitivity and its easy impedance load, Tethys will easily match with many high quality power amplifiers, giving to the listener an exciting and pleasurable sound experience.

Drivers	Ciare 30cm woofer, Ciare 13cm midrange, B&G planar tweeter
Frequency Response	27Hz – 20kHz
Sensitivity	90dB
Recommended Amplification	50W – 250W
Nominal Impedance	8 ohms
Crossover	450Hz, 5kHz
Dimensions (W x H x D)	16.4 x 51.1 x 18.8 inches / 410 x 1300 x 480 mm
Weight	132lbs / 60kgs
Available Finishes	Waxed Baltic Birch plywood and acrylic

SoundLab

Utah, USA
www.soundlab-speakers.com

SoundLab manufactures truly full-range electrostatic speakers. The distinctive characteristic of the panels is that the full audio spectrum emanates from every point on the same radiating surface, ensuring perfect time alignment, resulting in life-size pin-point imaging. Five panel sizes are offered that can be mounted in four different enclosure designs: the Ultimate, Majestic, Audiophile and Millennium series. A wide choice of grill fabrics, woods are offered, permitting an optimum match to virtually any sound room.

Majestic 945

Electrostatic Speaker
$36,883 (pair)

The Majestic 945 represents Soundlabs' best effort in the design of a truly full-spectrum electrostatic speaker. Frequency response covers the full audio spectrum without employing multiple drivers, such as woofers and tweeters, which create phase distortion resulting in compromised imaging. All sound energy emanates from a single very low-mass film, giving perfect time-alignment of all frequencies, thus providing optimum imaging and life-like clarity. Furthermore, the full audio spectrum is represented over the entire dispersion angle, permitting listening locations well off of the center axis of the speaker with no penalty in sound quality or imaging.

Frequency Response	24Hz - Ultrasonics
Power	50W – 600W
Nominal Impedance	8 ohms
Sensitivity	89dB
Horizontal Dispersion	45 degrees
Vertical Dispersion	Projected field of panel height
Power Supply	117VAC – 230VAC
Controls	High frequency, Mid frequency, Bass level, D.C. Bias
Dimensions (W x H x D)	39.9 x 104 x 8.3 inches / 1013 x 2642 x 211 mm
Weight	216lbs / 97.9kgs
Available Finishes	Medium Oak and other finishes

Surreal Sound

Virginia, USA
www.surreal-sound.net

Surreal Sound Audio, based in Chesterfield, VA is home to a new line of hand-crafted luxury speakers. You may never have heard bass as well defined and dynamic as with their full range speaker system. The midrange presentation is smooth and realistic, soul soothing as music should be. The high frequencies when properly done as by them, locate and stabilize the soundstage placing instruments in a virtual 3D world before you. You will always have concert quality sound without ever having to leave the comfort of your home.

Fifth Row

Loudspeaker
$27,500* (pair)

Surreal Sound's flagship speaker, the Fifth Row is named to replicate what they consider to be the best seat in the concert hall, Fifth Row Center. The cabinet design was twelve years in the making. The sub-woofer section holds six ten inch custom designed drivers, the midrange driver has a custom first order crossover at 125Hz and the tweeter is again customized to crossover at 3200Hz. Sub-woofer amps are included. The small footprint accommodates any medium or larger room. Best speaker, hand built for the discriminating listener.
*Price may vary.

Drivers	6 x 10 inch subwoofers
Frequency Response	20Hz – 20kHz
Crossover Frequency	125Hz
Nominal Impedance	8 ohms
Sensitivity	95dB
Recommended Amplification	8W - 35W
Options	Custom drivers available
Dimensions (W x H x D)	12 x 42 x 18 inches / 305 x 1066 x 457 mm
Weight	125lbs / 57kgs
Available Finishes	Custom finishes and colors

SVS

Ohio, USA
www.svsound.com

SVS is a global leader in high performance home audio products; designing, manufacturing and selling subwoofers and speakers through its Internet direct business. Founded in 1998 by a group of audio enthusiasts seeking to develop an alternative to traditional audio manufacturers, SVS was forged out of a passion for the science of sound married to a disruptive go-to-market strategy. Built on collective decades of experience, SVS is redefining performance and value for people who love music and sound.

SB-13 Ultra

Subwoofer
$1,599

Designed to combine best-in-class audio performance with advanced engineering, refined aesthetics, and easy room integration, the SB13-Ultra delivers size-no-object bass output from an elegant cabinet. Instead of merely adapting the famous 13.5-inch Ultra 13 driver for the SB13's new, more compact sealed enclosure, the engineers radically revised the motor geometry, added a unique, ultra-high-power aluminum voice coil, and developed a custom gap extension plate that increases linear stroke and reduces distortion. Dual linear-roll spiders, a stitched parabolic surround, and the ultra-light but rigid Rohacell composite cone material of the Ultra 13 driver result in an incredibly robust subwoofer with extreme excursion.

Driver	SVS 13.5 inch Ultra driver
Frequency Response	20Hz – 460Hz
Input Impedance	24k ohms
Inputs / Outputs	RCA, XLR
Amplifier	Class D STA-1000D
Amplifier Power	1000W rms
Peak Dynamic Power	3600W
Dimensions (W x H x D)	17.4 x 17.9 x 20.4 inches / 442 x 455 x 518 mm
Weight	92lbs / 41kgs
Available Finishes	Black

Tetra

Canada
www.tetraspeakers.com

At Tetra, they acknowledge that music is one of the most powerful forces for change on the planet and that the speaker is the most important component in the reproduction of music. For over ten years, they have insisted that the form and function of every Tetra must unite in a visually exciting and viscerally fulfilling experience. At Tetra, the philosophy is that every Tetra Listening Instrument must deliver the 'rush' of a physiological response to the music being played.

Tetra 333

3-Way Loudspeaker
$9,000 (pair)

Say hello to the Tetra 333 STACK. Comprised of the 222 bookshelf model for the 'tops' and the 111s for the 'bass bins', this building block approach allows for a smooth transition up the 'listening instrument' offering into what is ultimately a more easily attainable Tetra 3-way. So why delay? Get in touch with Tetra today and get yourself 'STACK'd... Tetra-style!

Drivers	5 inch glass fiber, 1 inch treated fabric, 8 inch polypropylene bass
Frequency Response	25Hz – 20kHz
Sensitivity	89dB
Nominal Impedance	8 ohms
Power Handling	250W
Dimensions (W x H x D)	12 x 43 x 17 inches / 300 x 1070 x 430 mm
Weight	56lbs / 25kgs
Available Finishes	Black, Red, Light wood

TR Studio

Poland
www.trstudio.com

Tomasz Rogula made his first loudspeakers in 1973. Since that time he has devoted his life to music, and the creation of audio systems for its reproduction. Working over the years with nearly 50,000 people visiting and recording at the internationally renowned recording studio TR Studios at Warsaw, it has been his goal to share with all - the meaning of "perfect sound." With modern research facilities available, the goal of Zeta Zero is to achieve a sound quality almost indiscernible from perfection for the average listener.

Zeta Zero - Venus Picolla

Loudspeaker

€22,990 (pair)

When you are looking for an undisturbed harmony of shape to match the beauty of your home interior design, the Zeta Zero Venus Picolla with its unique design is the perfect solution. Its precise spatiality eliminates the need to have an unsightly centre speaker in your home theatre installation. Venus Picolla speakers offer very high power, but in a small package with a very refined performance. The extreme dynamics of their system can generate a sound pressure in excess of ≈ 130 dB peak SPL - the same level you experience during a live rock concert.

Loudspeaker Type	Passive
Frequency Response	26Hz - 50kHz
Power	300W
Nominal Impedance	4 ohms
Sensitivity	>130dB
Recommended Amplification	8W - 1000W
Binding Terminals	Single / Bi-wiring
Dimensions (WxHxD)	18.5 x 42.5 x 18.8 inches / 470 x 1080 x 477 mm
Weight	143lbs / 65kgs
Available Finishes	Any. Black, Sahara-Sand, Mahogany-Brown, Transparent

Waterfall Audio

France
www.waterfallaudio.com

Waterfall Audio was established in 1996 by Cedric Aubriot, a pioneer in the Audio business. Truly unique designs, special attention to details and high-end sound performance are the key factors to their success. Innovation is their tradition, patented technologies guarantee perfect sound reproduction throughout their range of products. Luxury handcraftsmanship from France throughout Waterfall's assembly lines. Waterfall is now distributed in over 35 countries and has received many awards worldwide for design and acoustic excellence.

Niagara

3-Way Loudspeaker
$39,500 (pair)

Just imagine a loudspeaker and Home-Cinema system that will enhance your way of life! Dramatic style and luxury at every level! Stunning 47" towers of exclusive diamond glass, a specially engineered horn tweeter, solid Aluminum components and some parts finished in hand stitched Nappa leather. Niagara is a sophisticated blend of craftsmanship and innovative technology, delivering a rich natural and open sound beyond expectations. This unique design is the response to a need for ultra high-end audio speakers with cutting edge design. The Niagara speaker provides formidable performance with unequalled elegance and discretion.

Drivers	Neodymium Horn tweeter, 7 inch bass midrange, 8 inch ULF passive driver
Recommended Amplification	60W – 250W
Peak Power	600W
Nominal Impedance	4 ohms; 8 ohms
Sensitivity	89dB
Frequency Response	36Hz – 28kHz
Dimensions (W x H x D)	11.8 x 47.3 x 12.6 inches / 300 x 1200 x 320 mm
Weight	132lbs / 60kgs
Available Finishes	Clear Diamond Glass

Westone Music

Colorado, USA
www.westonemusicproducts.com

After many decades of in-ear and acoustic innovation, Westone has developed a sizable portfolio of exclusive technologies and incorporates only the most premium materials and optimal processes into all its products. Just as the best musical instruments are those constructed by hand, so are the best earphones which reproduce your music. Meticulously built by a staff of dedicated artisans and lab techs, Westone's USA products exhibit a level of craftsmanship that truly embodies America's rich history of handmade artistry.

ADV Alpha (Adventure Series)

Cross-Over Earphone
$249.99

The world's first cross-over earphone, the Adventure Series Alpha is ready for anywhere life's soundtrack will take you. The magnesium unibody design is durable and stylish while the AWACS Reflective cable is replaceable and boasts a 3 button control system and mic. You'll also be ready for adverse weather conditions. And thanks to the Active Fit System, your Adventure Series ALPHA is comfortable and will stay in place for all-day comfort and performance.

Driver	6.5 mm Micro Driver
External Protection	Weather resistant IPX-3
Body	Magnesium unibody and Aluminum faceplate
Features	Adventure Warning and Control System
Compatibility	iPod / iPhone / iPad
Sensitivity	97dB
Frequency Response	20Hz - 18kHz
Nominal Impedance	21 ohms
Cable	AWACS Reflective
Available Finishes	Black

Wisdom Audio

Nevada, USA
www.wisdomaudio.com

Wisdom Audio was founded in 1996 with one goal in mind, to create the world's finest loudspeakers unbound by limitations of current technologies and past design. Wisdom's pioneering use of large scale thin film planar magnetic transducers, electronic crossovers in place of passive designs, and innovative woofer designs were all developed to solve problems inherent in the limitations of traditional designs and how they interact with the listening room. The result is an experience like no other.

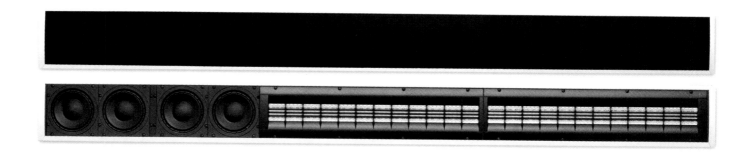

L75

Free-standing Speaker
$18,700 (pair)

For those customers seeking a flexible application, high performance architectural loudspeaker, the L75 is the solution. Merging the benefits of the superior qualities of a true PMD line source and outstanding low frequency performance, the L75 defies categorization. The L75's 48-inch tall planar magnetic line source makes it capable of unprecedented dynamics, low distortion, high resolution, faithful timbre reproduction, three dimensional sound staging and freedom from compression. This system should be compared with the best freestanding systems and needs to be auditioned to be fully appreciated.

Frequency Response	40Hz - 20kHz
Nominal Impedance (Planar)	4.5 ohms
Nominal Impedance (Woofer)	3 ohms
Sensitivity (Planar)	95dB
Sensitivity (Woofer)	91dB
Dimensions (W x H x D)	7.1 x 76.1 x 6.3 inches / 181 x 1930 x 160 mm
Weight	79lbs / 36kgs
Available Finishes	Black

YG Acoustics

Colorado, USA
www.yg-acoustics.com

Established in 2002, YG Acoustics is a leader in high-end loudspeaker engineering. YG Acoustics was founded on the principles of correcting the inherent weaknesses in the design process of loudspeakers. As a result the company produced unique software for speaker design that would factor in the all important aspects of phase and amplitude response, producing simultaneous computation for perfect linearity in both domains. Known as DualCoherent™, this technology is now used in all YG Acoustics' loudspeakers.

Sonja 1.3
Loudspeaker
$106,800

Sonja is YG Acoustics' flagship, and represents everything that the world knows about speaker-design. Driven by innovative applications of acoustical engineering and world-class precision-manufacturing, Sonja pushes the edge of the art. Marrying these elements with stunning industrial design, results in a loudspeaker which is transformative in both performance and sheer visual beauty. Sonja - unparalleled sonics, timelessly elegant form, pure seduction.

Drivers	BilletCore™, ForgeCore™
Frequency Response	20Hz - 40kHz
Crossover	65Hz, 1.75kHz
Sensitivity	88dB
Nominal Impedance	4 ohms
Minimum Impedance	3 ohms
Dimensions (W x H x D)	17 x 70 x 28 inches / 432 x 1778 x 711 mm
Weight	506lbs / 230kgs
Available Finishes	Black

60 Aaron & Sovereign Germany	**61** Aaron & Sovereign Germany	**62** Aaron & Sovereign Germany	**63** Aaron & Sovereign Germany	**64** Abbingdon Music Research United Kingdom	**65** Abbingdon Music Research United Kingdom
66 Aesthetix California, USA	**67** Antelope Audio California, USA	**68** Arcam United Kingdom	**69** Arcam United Kingdom	**70** ASR Audio Systems Germany	**71** Audioengine Hong Kong
72 Audioengine Hong Kong	**73** Audio Research Minnesota, USA	**74** Aurorasound Japan	**75** Canor Audio Slovakia	**76** Canor Audio Slovakia	**77** Channel D USA
78 Classe Canada	**79** Coffman Labs Portland, USA	**80** Dan D'Agostino Connecticut, USA	**81** Devialet France	**82** Eryk S Concept Poland	**83** Erzitech Audio Slovenia
84 Fosgate California, USA	**85** Gato Audio Denmark	**86** Hagerman Hawaii, USA	**87** ISOL-8 Teknologies United Kingdom	**88** Kharma Netherlands	**89** KingRex Technology Taiwan
90 KingRex Technology Taiwan	**91** KR Audio Czech Republic	**92** Krell Connecticut, USA	**93** Lehmannaudio Germany	**94** LessLoss Lithuania	**95** Linar Audio Canada
96 Lynx Studio California, USA	**97** McIntosh New York, USA	**98** Musical Surroundings California, USA	**99** Nagra Switzerland	**100** NAT Audio Serbia	**101** Norma Audio Italy
102 Plitron Manufacturing Canada	**103** Pureaudio New Zealand	**104** Redgum Audio Australia	**105** Roksan Audio United Kingdom	**106** Roksan Audio United Kingdom	**107** Sanders Sound Systems Colorado, USA
108 Soulution Switzerland	**109** Stahl~Tek Texas, USA	**110** Spiritual Audio California, USA	**111** Sutra Italy	**112** Synthesis Italy	**113** Synthesis Italy
114 Synthesis Italy	**115** Tangent Audio Denmark	**116** TBI Audio Systems United Kingdom	**117** Teddy Pardo Isreal	**118** Teddy Pardo Isreal	**119** Trinity Electronic Design Germany
120 Trinity Electronic Design Germany	**121** Viola Audio Laboratories Connecticut, USA	**122** Wadia New York USA	**123** Wadia New York USA	**124** Wyred 4 Sound California, USA	**125** Zesto Audio California, USA

Aaron & Sovereign

Germany
www.aaron-amplifiers.com

Their company, the high-end consumer Electronics Distribution Company m.b.H, has been a think tank, as well as a manufacturing and distribution company for more than a quarter century, founded in 1986 by Marita and Thomas Hoehne. Their vision of entertainment systems that could faithfully reproduce music was the company's declared objective. Their focus on the essentials, the heart and soul of high-end hi-fi equipment and amplifiers, was received the special attention of the Hoehne family.

Aaron XX

Stereo Integrated Amplifier
$2,990

The XX is the Anniversary Limited Edition by Aaron. Also called "Double X", the name comes from the number 20 in Roman numerals and recalls the year 1989, the year of the AARON launch party. It is designed as an amplifier with the very best price-sound ratio. Like all of their amplifiers, the XX is also lovingly manufactured by hand. The XX even has a preamp output that allows you to use another stereo amp for bi-amping your speakers, if they are designed for this mode. The XX will amaze you. Everyday.

Output Power	270W
Maximum Current	16A
Harmonic Distortion	0.014%
Intermodulation Distortion	0.017%
Frequency Response	DC – 120kHz
Input Impedance	47k ohms
Output Impedance	8 ohms
Gain	40dB
Power Consumption (Idle)	6VA
Dimensions (W x H x D)	17.3 x 4.3 x 14.7 inches / 440 x 108 x 375 mm

Aaron & Sovereign

Germany
www.aaron-amplifiers.com

Their company, the high-end consumer Electronics Distribution Company m.b.H, has been a think tank, as well as a manufacturing and distribution company for more than a quarter century, founded in 1986 by Marita and Thomas Hoehne. Their vision of entertainment systems that could faithfully reproduce music was the company's declared objective. Their focus on the essentials, the heart and soul of high-end hi-fi equipment and amplifiers, was received the special attention of the Hoehne family.

Aaron No.3 Millenium

Power Amplifier
$3,590

The No.3 Millennium is an amplifier that carries your speakers to sonic excellence. In particular, as the owner of very high quality, sophisticated or powerful speaker systems, you'll be thrilled. The almost legendary stability of the No.3 Millennium amplifier, even on speakers with critical impedance processes, or on systems that are difficult to operate, is world-famous among dedicated high-end fans. The special circuit layout and the use of selected components on the power modules, coupled with a generously dimensioned power supply, ensures an impressive music experience.

Output Power	350W
Maximum Current	20A
Harmonic Distortion	0.0087%
Intermodulation Distortion	0.014%
Frequency Response	DC – 160kHz
Input Impedance	47k ohms
Output Impedance	8 ohms
Gain	34dB
Power Consumption (Idle)	7VA
Dimensions (W x H x D)	17.3 x 4.3 x 14.7 inches / 440 x 108 x 375 mm
Available Finishes	Silver, Black

Aaron & Sovereign

Germany
www.sovereign-amplifiers.com

Their company, the high-end consumer Electronics Distribution Company m.b.H, has been a think tank, as well as a manufacturing and distribution company for more than a quarter century, founded in 1986 by Marita and Thomas Hoehne. Their vision of entertainment systems that could faithfully reproduce music was the company's declared objective. Their focus on the essentials, the heart and soul of high-end hi-fi equipment and amplifiers, was received the special attention of the Hoehne family.

Sovereign Director
Preamplifier
$15,990

The ultimate high-end preamplifier for the serious stereo lover and analog music reproduction enthusiast. An absolute precision instrument that combines luxury, elegance, tradition and state-of-the-art technology to enthrall you: a connoisseur of music. Designed and created to experience emotional events in the purest fashion. It will touch you, impress you and fascinate you again and again through the music you love most. Experience your music the way you always wanted to: totally natural.

Output Signal	16V
Harmonic Distortion	0.0085%
Intermodulation Distortion	0.0065%
Frequency Response	3Hz - 120kHz
Gain	45dB
Input Impedance	47k ohms
Output Impedance	1 ohm
Power Consumption (Idle)	5VA
Dimensions (W x H x D)	19 x 4.7 x 11.8 inches / 480 x 120 x 300 mm
Weight	33lbs / 15kgs

Aaron & Sovereign

Germany
www.sovereign-amplifiers.com

Their company, the high-end consumer Electronics Distribution Company m.b.H, has been a think tank, as well as a manufacturing and distribution company for more than a quarter century, founded in 1986 by Marita and Thomas Hoehne. Their vision of entertainment systems that could faithfully reproduce music was the company's declared objective. Their focus on the essentials, the heart and soul of high-end hi-fi equipment and amplifiers, was received the special attention of the Hoehne family.

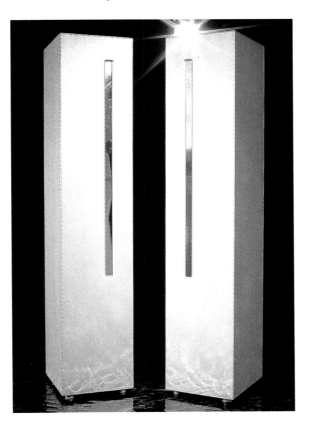

The Sovereign
Monoaural Power Amplifier
$179,900

The measure of all things. The best high end amplifier in the world. The world's biggest power amplifiers. Music adventures in live quality. Way ahead of all high end amplifiers by at least a decade. The Sovereign is their dream in the shape and form of an amplifier. Even today the roughly one and a half metre high amplifier twin towers stand for a developing process unchallenged in any regard from its beginning to its end. "No Limits" was the challenge when the company started with developing the "best possible amplifier" many years back. Today, and from now on, this also applies to the further development of "The Sovereign".

Operation	3 Phase
Harmonic Distortion	0.009%
Intermodulation Distortion	0.010%
Frequency Response	DC - 180kHz
Input Impedance	47k ohms
Gain	50x = 34dB
Damping Factor	900
Power Consumption (Idle)	20VA
Slew Rate	35V/microseconds
Dimensions (W x H x D)	11.8 x 59.8 x 13.7 inches / 300 x 1520 x 350 mm
Weight	306lbs / 139kgs

Abbingdon Music Research

United Kingdom
www.amr-audio.co.uk

Abbingdon Music Research is a high-end manufacturer of some of the finest audio components. The 77 Reference Class, the 777 Premier Class and its range of accessories have all garnered critical acclaim. Since the CD-77 which was first launched in 2006 through to the latest DP-777, all boast ground-up technology. AMR's class re-defining audio components have justly forged an enviable reputation to move the listener with the stirring emotion, verve and faithfulness of the original performance.

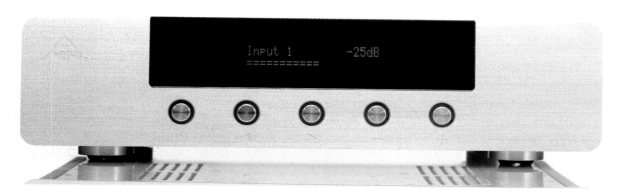

AM-777

Pre-Main Amplifier
$4,999

The AM-777 Premier Class Pre-Main Amplifier breaks with tradition and possesses eye-catching features not seen in other amplifiers at any price. At the heart, based upon AMR's OptiGain® circuitry, it has the same unique sonic DNA as its larger sibling: the AM-77. The AM-777 is a serious amplifier that, first and foremost, is devoted to reproducing music that is emotionally stirring, reminiscent of the best valve amplifiers. What makes the AM-777 even more enticing is that it has onboard, an advanced, sonically superior USB digital-to-analogue converter optimised by AMR for the highest quality signal transfer.

Inputs	RCA, XLR, USB
Frequency Response	10Hz - 30kHz
Signal-to-Noise Ratio	>100dB
Harmonic Distortion	<0.3%
Power Consumption (Standby)	<1W
Output Impedance	0.33 ohms
Input Impedance	20k ohms
Operational Modes	Pre-main Amplifier, Power Amplifier, Monoblock Option
Dimensions (W x H x D)	17.7 x 4.7 x 14.6 inches / 449.6 x 119.4 x 370.8 mm
Weight	31lbs / 14kgs
Available Finishes	Silver, Black

Abbingdon Music Research

United Kingdom
www.amr-audio.co.uk

Abbingdon Music Research is a high-end manufacturer of some of the finest audio components. The 77 Reference Class, the 777 Premier Class and its range of accessories have all garnered critical acclaim. Since the CD-77 which was first launched in 2006 through to the latest DP-777, all boast ground-up technology. AMR's class re-defining audio components have justly forged an enviable reputation to move the listener with the stirring emotion, verve and faithfulness of the original performance.

DP-777

Digital Processor
$4,999

High-Definition digital audio is redefining the quality of recorded audio for the 21st Century. Having created some of the most natural and captivating music sources on the CD standard, AMR has turned its attention to High-Definition audio and in its own imitable way, taken the road less traveled. Their first objective was to deliver the maximum quality of music from the existent libraries of CD standard audio equally balanced with the requirement to also deliver the maximum music quality from the new HD audio formats. The result is the DP-777 Digital Processor.

Audio Inputs	XLR, RCA, USB
Output Voltage	>2V
Frequency Response	20Hz – 20kHz
Signal-to-Noise Ratio	>100dB
Harmonic Distortion	<0.3%
Dynamic Range	>90dB
Power Consumption (Standby)	<1W
Rated Voltage	115VAC – 230VAC
Dimensions (W x H x D)	17.7 x 4.7 x 14.6 inches / 449.6 x 119.4 x 370.8 mm
Weight	25lbs / 11.5kgs
Available Finishes	Silver, Black

Aesthetix

California, USA
www.aesthetix.net

Aesthetix is the brain child of engineer and music lover Jim White. Avid concert-going from the age of 10 were the early hints of what would become a life consuming obsession with music and vinyl playback. Aesthetix products are meticulously hand assembled by the talented factory technicians. Structural design is given more attention than competing products. Constrained-layer damping, transformer isolation, and other techniques are employed to lower mechanical noise to a minimum.

Atlas Signature Stereo

Power Amplifier
$10,000

Atlas amplifiers are capable of driving virtually any high quality speaker. An innovative hybrid design incorporates a bipolar output stage, bipolar driver stage, and vacuum tube input gain stage. They stand alone as the only hybrid amplifiers with zero feedback, essential for maintaining harmonic integrity, air, space and coherence. The Atlas provides your choice of full range or high-pass inputs. This unique built-in filter is ideal for speakers featuring powered woofers, such as Vandersteen, or for audio and home cinema systems using outboard powered sub-woofers.

Power Output	2 x 200W @ 8 ohms, 2 x 400W @ 4 ohms
Input Sensitivity	60mV@1W
Input Impedance	470k ohms
Output Impedance	0.25 ohms
Signal-to-Noise Ratio	120dB
Frequency Response	4Hz – 150kHz
Harmonic Distortion	<1%
Power Consumption (Idle)	100W
Dimensions (W x H x D)	18 x 8 x 19 inches / 457.2 x 203.2 x 482.6 mm
Weight	70lbs / 31.7kgs
Available Finishes	Black, Silver

Antelope Audio

California, USA
www.antelopeaudio.com

Antelope is the brainchild of Igor Levin, best known for creating the legendary AardSync. With over twenty years' experience and an unparalleled understanding of digital audio, Igor continues to establish new standards in quality and performance. Antelope was founded to take audio recording to the next level, providing digital technology that transcends all expectations. Antelope allows you to harness the power of digital audio without sacrificing the warmth and fullness typically associated with analog gear.

Rubicon
AD/DA Preamplifier &
Headphone Amplifier
$40,000

Rubicon is the World's first converter, phono stage preamp and headphone amplifier with an integrated atomic clock. DLNA streaming, the JFET phono preamp, the DAC and the ultra-high sample rate A/D conversion, together with the high resolution USB recording capability provide the users with the sublime experience of digitizing their favorite tracks, still keeping the depth and the warmth of the original analog recordings. With Antelope, it's truly possible to have the best of both worlds.

Atomic Clock	10M Rubidium
DAC / ADC	384kHz
Analog Outputs	Balanced output XLR, Unbalanced output on RCA
Digital Outputs	S/PDIF De-jittered outputs, Hi-Speed USB
DLNA Streaming	Through Ethernet
Headphone Amps	Dual Stage, Ultra-linear
Volume Attenuator	Matched to 0/.05dB for all volume levels
Streaming	Custom USB 2.0 chip up to 480M bits/384kHz
Jitter Management	64-bit acoustically Focused Clocking
Dimensions (W x H x D)	24 x 10 x 24 inches / 610 x 254 x 610 mm
Weight	60lbs / 27 kgs

Arcam

United Kingdom
www.arcam.co.uk

Arcam first began building sound reproduction equipment in 1972, whilst its founders were still science and engineering students at Cambridge University. Arcam exists to bring the highest fidelity reproduction of music and movies into people's homes. They are committed to engineering products that deliver a level of audio performance that is so convincing and lifelike that it connects you straight to the emotional power of the music. If you love music and movies then you need to experience the magic that an Arcam-based system can produce.

FMJ A19

Integrated Amplifier
£650

While the world of music is increasingly a digital one, analog sources are not forgotten. The continued popularity of vinyl has led Arcam to equip the A19 with a new ultra-low noise moving magnet (MM) phono stage that allows the music enthusiast to enjoy their vinyl collection fresh. While the use of high performance headphones has grown dramatically of late, little effort has been made to improve the headphone sections of the amplifiers that drive them. The A19 addresses this by employing a completely new circuit that can deliver a level of sound quality that will thrill even the most demanding listener.

Power Output	50W
Frequency Response	20Hz – 20kHz
Harmonic Distortion	0.003%
Signal-to-Noise Ratio	80dB - 105dB
Input Impedance	10k ohms
Output Impedance	1 ohm – 230k ohms
Power Consumption (maximum)	350W
Load Range	16 ohms – 2k ohms
Dimensions (W x H x D)	17 x 3.3 x 10.8 inches / 432 x 85 x 275 mm
Weight	18.7lbs / 8.5kgs
Available Finish	Black

Arcam

United Kingdom
www.arcam.co.uk

Arcam first began building sound reproduction equipment in 1972, whilst its founders were still science and engineering students at Cambridge University. Arcam exists to bring the highest fidelity reproduction of music and movies into people's homes. They are committed to engineering products that deliver a level of audio performance that is so convincing and lifelike that it connects you straight to the emotional power of the music. If you love music and movies then you need to experience the magic that an Arcam-based system can produce.

r-Series irDAC

DAC
£400

The irDAC is so called because it features infra-red remote control, owes much to the development programme behind the reference D33 DAC. The irDAC comes packed with cutting edge technology that brings high-end performance to a wide array of sources. It's outstanding sound quality will deliver sheer musical enjoyment for all music lovers. The irDAC uses the outstanding Burr Brown 1796 DAC and 8 separately regulated power supplies to ensure class leading performance that are unmatched in the irDAC's price category. The irDAC is designed to be the heart of a digital system and can be connected to a host of different types of digital sources and connections.

DAC	Burr-Brown/TI PCM1796
Frequency Response	10Hz – 20kHz
Harmonic Distortion	0.002%
Signal-to-Noise Ratio	112dB
Maximum Output Level	2.2Vrms
Sampling Rate	Up to 192kHz
Bit Depth	16-bit; 24-bit
Power Requirements	12V DC, 1.5A
Accessories	PSU, Remote, Cables
Dimensions (W x H x D)	7.4 x 1.7 x 4.72 inches / 190 x 44 x 120 mm
Weight	2.2lbs / 1kg

ASR Audio Systems

Germany
www.asraudio.de

ASR audio systems since 1980, successfully builds high-end amplifiers. Frederick Schaefer, owner and director of development at ASR Audio Systems, has a clear idea of how an amplifier should sound: homogeneous and natural! With ASR electronics you can enjoy long relaxing music! The company remains loyal to its philosophy, to reproduce music as a complete experience. This synthesis will unfold the aliveness of music, the listening room will become transparent and allow the reproduction of delicate music.

Emitter II Exclusive

Amplifier
$27,700*

The Emitter II Exclusive (Dual Mono) sets another milestone in the consistent further development of the Emitter II. All possible improvements of the Emitter II are integrated into this version. The sound picture is very fast and dynamic, and the audio solution is increased. Furthermore, the sound is quiescent and even more relaxing, smooth and homogeneous. Through its extremely high speed, the Emitter II Exclusive reproduces authentically the three dimensions of good recordings into the listening room. *Price is inclusive of power supply.*

Output Power	2 x 280W @ 8 ohms, 2 x 490W @ 4 ohms, 2 x 900W @ 2 ohms
Frequency Response	1Hz - 100kHz
Signal-to-Noise Ratio	>86dB
Distortion	<0.01%
Inputs/Outputs	RCA, XLR
Power Consumption (Standby)	20VAC
Rise Time	<8 microseconds
Dimensions (W x H x D)	22.5 x 9 x 18.5 inches / 572 x 229 x 470 mm
Weight	104lbs / 47.2kgs
Available Finishes	Gold, Version Blue
US Distributor	www.musicalsounds.us

Audioengine

Hong Kong
www.audioengineusa.com

Audioengine products are based on custom designs with very few off-the-shelf parts. After years of building professional powered studio monitor speakers, Audioengine has taken their experience and created unique powered consumer speakers for your home and desktop. Now Audioengine is into it's 6th year and their goal remains the same - to give customers a high-end listening experience at affordable prices with easy to use, highquality products. Audioengine invites you to join the computer audio revolution!

Audioengine W3

Wireless Audio Adapter
$149 (pair)

Audioengine W3 Premium Wireless Audio Adapter - Play all your music wirelessly from any audio device or computer to your Audioengine powered speakers, stereo receiver, or powered subwoofer. The W3 consists of 2 parts, the "Sender" and "Receiver". The Sender transmits audio from your computer through USB audio or from any product with 3.5 mm mini-jack or RCA audio outputs. The other side of W3, the Receiver, connects audio to any product with mini-jack or RCA audio connectors. W3 provides CD quality HD stereo sound with no reduction in audio quality.

Converter	DAC: CS4344, ADC: CS5343
Inputs	USB audio mini-jack analog
Outputs	Mini-jack analog
Output Impedance	470 ohms
Signal-to-Noise Ratio	95dB
Harmonic Distortion	0.01%
Frequency Response	20Hz - 20kHz
Wireless Range	>100ft
Dimensions (W x H x D)	4 x 0.4 x1.2 inches / 100 x 10 x 30 mm
Weight	1lb / 0.45kgs

Audioengine

Hong Kong
www.audioengineusa.com

Audioengine products are based on custom designs with very few off-the-shelf parts. After years of building professional powered studio monitor speakers, Audioengine has taken their experience and created unique powered consumer speakers for your home and desktop. Now Audioengine is into it's 6th year and their goal remains the same - to give customers a high-end listening experience at affordable prices with easy to use, highquality products. Audioengine invites you to join the computer audio revolution!

Audioengine D1

DAC
$169

D1 is the perfect digital interface between your computer and music system and will improve the sound of all your music. The D1 accepts inputs from both USB and optical and has outputs for any audio system or headphones. Audioengine designers, with their attention to audio quality as well as visual aesthetics, have created a feast for both the eyes and ears. The rounded anodized aluminum case of the D1 reflects traditional Audioengine design. The Audioengine D1 continues to close the gap between your computer music and home hi-fi.

DAC	AKM4396
Optical Receiver	CS8426
DAC type	Dual Mode USB and Optical (SPDIF)
Inputs	USB / Optical (SPDIF)
Outputs	RCA stereo / 3.5 mm headphone
Frequency Response	10Hz - 25kHz
Signal-to-Noise Ratio	>110dB
Harmonic Distortion	<0.002%
Output Impedance	47 ohms RCA, 10 ohms headphone
Dimensions (W x H x D)	3.5 x 1 x 4 inches / 88.9 x 25.4 x 101.6 mm
Weight	1lbs / 0.5kgs

Audio Research

Minnesota, USA
www.audioresearch.com

Audio Research is one of the oldest continually operating manufacturers in American audio. The company was founded in 1970 in Minneapolis with one goal in mind: to advance the state-of-the-art in music reproduction and was instrumental in refocusing the audio industry on designing products for musical performance. Much more than a mere carriage-trade brand name, the Audio Research marque has come to represent stellar performance and lasting value for music lovers and audio enthusiasts everywhere.

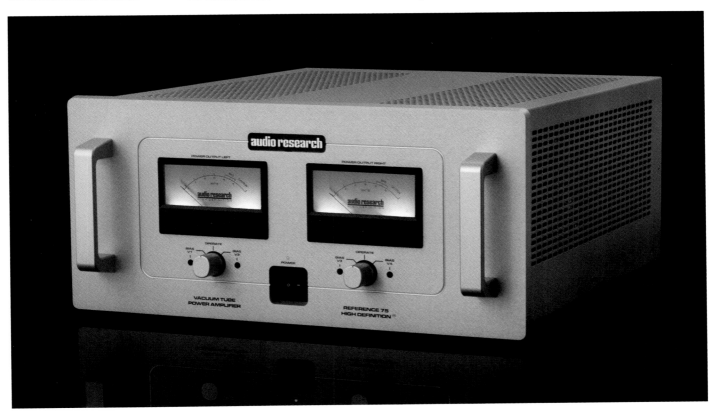

Reference 75

Vacuum Tube Power Amplifier
$9,000

The Reference 75 vacuum tube stereo amplifier has garnered universal praise during its short life in the Audio Research Reference series. Though rated at 75 watts per channel, this amplifier has the ability to drive a wide variety of speakers to bring music to life in your listening room. The signal path in the Reference 75 has been designed to be as short and clear as possible, providing the most direct link from source to speaker. With a complete lack of grain or artifice, the Reference 75 produces beautiful, palpable sound and elevates your entire music library to an entirely new experience.

Power Output	2 x 75W
Frequency Response	0.7Hz – 75kHz
Input Impedance	300k ohms
Polarity	Non-Inverting
Slew Rate	10V/microsecond
Rise Time	4.0 microseconds
Hum & Noise	< 0.06mV rms
Power Requirements	105VAC – 125VAC / 210VAC – 250VAC
Tubes Required	Matched pair KT120, 6H30 Driver
Dimensions (W x H x D)	19 x 8.7 x 19.5 inches / 483 x 222 x 495 mm
Weight	47lbs / 21.3kgs

Aurorasound

Japan
www.aurorasound.jp

Aurora is the sound studio of handmade amp builder in town and harbor sea, in Yokohama. With internal structure that is carefully handmade one by one, physical measurements and electrical, as well as design approach bold and faithful, the basic in-depth is assembled in final judgment by the ear. The idea, to produce the audio equipment as craftsman to create a musical instrument is turned philosophy wants to be a work rather than industrial products amplifier.

Vida

Phono Equalizer
$4,990

Across the entire frequency range, RIAA curve is compensated with constant condition. As a result, achieved very low distortion and high resolution in low frequency, massive mid range and very stable and bright high range. VIDA is designed to combine old technology and new state of the art semiconductor to create unique phono stage amplifier. The inductor is specially designed by Swedish company " Lundahl" for Aurorasound. Power supply is a separate chassis to secure very low signal to noise ratio with high level dynamic range. Old fashioned compact size wood case would be a good partner for your analog music enjoyment.

Inputs	1 x MM Gain 39dB 47k ohms, 1 x MC Gain 64dB
Outputs	Line Level, unbalanced RCA, optional XLR balanced
Frequency Response	10Hz – 20kHz
Harmonic Distortion	0.025%
Input Noise	-138dBV
Dimensions (WxHxD)	10.2 x 4 x 9.8 inches / 260 x 100 x 250 mm
Power Supply	100V - 120V / 220V - 240V
Weight	6.6lbs / 3kg

Canor Audio

Slovakia
www.canor-audio.com

Canor is a new star, on the high-end tube sky, producing the ultimate design "cost no object" electronics. The products incorporate internally the best of classic tube technology. The products are super heavy inside with lots of the very best components, but on the outside has a very clean and functional design. Canor has been developing and manufacturing high-end audio products since the 80s. All technological processes in the production lead to the only aim, to manufacture high-quality and reliable products.

TP134
Integrated Amplifier
$3,990

The Canor TP134 is an integrated all-tube amplifier with EL34 high-power tubes in ultralinear connection, which operates in pure class A up to 10W per channel. Massive welded metal structure eliminates mechanical vibrations. The TP134 is the first amplifier in their product line in which for volume control a relay attenuator is being used, instead of a standard potentiometer. One of the most important parts of a tube amplifier, an output transformer, has several bifilar windings which are lined up in sections to maximise power bandwidth.

Tube Complement	4 x 12AT7(ECC81), 4 x EL34
Output Power	2 x 35W
Frequency Response	20Hz - 20kHz
Input Impedance	30k ohms
Inputs	5
Harmonic Distortion	<0.15%
Signal-to-Noise Ratio	93dB
Power	230V
Dimensions (W x H x D)	17.1 x 6.7 x 15.3 inches / 435 x 170 x 390 mm
Weight	48.5lbs / 22kgs
Available Finishes	Silver, Black

Canor Audio

Slovakia
www.canor-audio.com

Canor is a new star, on the high-end tube sky, producing the ultimate design "cost no object" electronics. The products incorporate internally the best of classic tube technology. The products are super heavy inside with lots of the very best components, but on the outside has a very clean and functional design. Canor has been developing and manufacturing high-end audio products since the 80s. All technological processes in the production lead to the only aim, to manufacture high-quality and reliable products.

TP10

Headphone Amplifier
$1,050

The Canor TP10 is a headphone tube amplifier of hybrid structure, and it is a direct successor to the highly popular and successful SH-1 model. The new model has a completely redesigned and improved power supply. The amplifier uses superior operational amplifiers (Burr-Brown), and top-echelon polypropylene capacitors. The tube amplifier is powered from an external adaptor. External adaptor significantly increases signal-to-noise ratio which is considered, together with musical expression of the amplifier, as one of the most important qualities of the amplifier.

Output Power	400mW
Gain	10.5dB
Input Impedance	47k ohms
Harmonic Distortion	<0.03%
Frequency Response	30Hz – 20kHz
Signal-to-Noise Ratio	>97dB
Headphone Jack	3-pole, 6.3 mm
Power Supply	16V
Dimensions (W x H x D)	8.2 x 3.4 x 11.6 inches / 210 x 88 x 295 mm
Weight	6.6lbs / 3kgs
Available Finishes	Silver, Black

Channel D

New Jersey, USA
www.channld.com

Channel D is the creator of the award-winning Pure Music® and Pure Vinyl™ high resolution digital music player, and vinyl transcription software for Apple Macintosh computers. Channel D's product line leverages their unique expertise in both audio software and audio hardware, linking vinyl playback with high resolution digital music reproduction. The Seta® line of ultra-wide bandwidth "flat" phono preamplifiers, perfect for use with Pure Vinyl software, includes the Stereophile Editors' Choice Class A - rated Seta Model L.

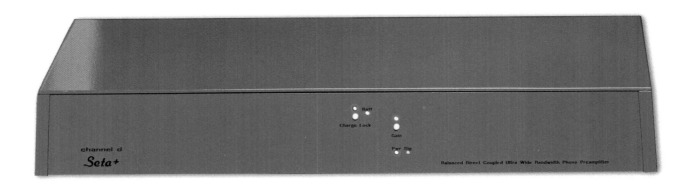

Seta Model L

Pre-Amplifier
$5,899

Seta phonograph preamplifiers provide the users of Channel D Pure Vinyl software with an end-to-end route to the superb performance attainable through "flat" vinyl reproduction. Seta offers a fully balanced low-noise differential circuit design, with unprecedented signal bandwidth. The result is stunning improvements in smoothness, clarity, definition and 3-D imaging. With Pure Vinyl applying the RIAA phono correction curve, the strengths of the latest 21st century cutting-edge analog and digital technologies are brought together, delivering superb, high definition, transparent vinyl playback.

Voltage Gain	43dB – 53dB
Input Impedance	2k ohms
Frequency Response	DC to >5MHz
Harmonic Distortion	<0.001%
Power Consumption (Standby)	<5W
Output Impedance	<40 ohms
Phase Shift	<1 degree
Propagation Delay	<60 nanoseconds
Inputs & Outputs	Neutrik, XLR, RCA, WBT
Dimensions (W x H x D)	12 x 2 x 7 inches / 305 x 51 x 178 mm
Weight	6.5lbs / 2.9kgs

Classe

Canada
www.classeaudio.com

Chairman, Mike Viglas, at the helm of Classé for more than thirty years, began his involvement as an audiophile customer who fell in love with the original Classé amplifiers, designed by founder David Reich. In 2001, Classé became part of the B&W Group. Its components are developed by specialists. The manufacturing employs both automatic and manual assembly procedures. For those who invest in it, the entertainment experience can be truly breathtaking. Every Classé component is designed and built to last a lifetime.

CP-800

Stereo Pre-Amp / Processor
$5,000

The CP-800 combines the functionality of a conventional analog preamplifier with that of a multi-input DAC having digital processing features. The single-box approach eliminates unnecessary chassis, circuitry and cables, which reduces cost, shortens the signal path and preserves signal quality. The CP-800's processing capabilities allow performance-enhancing features like parametric EQ, tone control and bass management to be accomplished in the digital domain, preserving the resolution of signals in a way not possible with conventional analog-based systems.

Frequency Response	8Hz - 20kHz
Distortion	0.002%
Gain	+14dB
Input Impedance	50k ohms (balanced); 100k ohms (single-ended)
Output Impedance	300 ohms (balanced); 100 ohms (single-ended)
Signal-to-Noise Ratio	104dB
Crosstalk	>-130dB
Power Consumption (standby)	<1W
Dimensions (W x H x D)	17.5 x 4.8 x 17.5 inches / 445 x 121 x 445 mm
Weight	23lbs / 10kgs

Coffman Labs

Portland, USA
www.coffmanlabs.com

As a classical violinist, physicist, Damon Coffman founded Coffman Labs with one goal: creating products that reproduce the natural sound experienced during live performances. Coffman Labs builds all-analog, vacuum tube-based products including the award-winning G1-A phono/line preamplifier and the new H1-A headphone amplifier. Products are all hand-built in Oregon, USA, using point-to-point wiring and custom parts, resulting in exceptional sound and heritage durability.

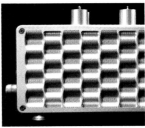

G1-A

Vacuum Tube Pre-Amplifier
$5,795

In development for over 3 years, the award-winning Coffman Labs G1-A is a limited edition vacuum tube-based preamplifier with an external power supply. Features include a Class A MM/MC phono stage, linestage, plus an adjustable headphone output to accommodate many headphone types. To achieve best quality, and organic sound, the G1-A contains rare and custom parts like military-grade aircraft switches and oil-in-paper capacitors. Custom-made chokes replace resistors in the signal path. There are no circuit boards: all wiring is point-to-point by hand.

Frequency Response	20Hz – 100kHz
Signal-to-Noise Ratio	85dB
Gain	14dB, 36dB, 52dB, 60.5dB, 76dB
Inputs / Outputs	REC, Phono, Mute, S1, S2, S3 / Inverted, Non-Inverted (Hi/Low)
Power Supply	110V -120V or 220V - 240V
Options	MM, MC
Dimensions (W x H x D)	9.5 x 7.7 x 12.5 inches / 241 x 196 x 318 mm

Dan D'Agostino

Connecticut, USA
www.dandagostino.com

No name is more closely associated with high-end audio amplifiers than that of Dan D'Agostino. During his career of more than 30 years, D'Agostino pioneered countless advances in the design of amplifiers, preamplifiers, CD players, and surround-sound processors. He is known as the audio industry's most passionate promoter of high quality, high-powered amplification. The new Dan D'Agostino products also reflect his concern for the environment. Makers of the world's finest loudspeakers rely on D'Agostino-designed amplifiers for their most important demonstrations.

Momentum

Monoblock Amplifier

$55,000

The Momentum monoblock amplifier is the ultimate expression of legendary audio designer Dan D'Agostino's passion for music. A distinctive meter, inspired by the styling of fine Swiss watches and bearing D'Agostino's signature, allows monitoring of power output. It delivers an incredible 300 watts into 8 ohms, but thanks to its highly efficient copper heat sinks with cooling venturis, the chassis is just 5 inches high and consumes less than 1 watt of power at idle. Each Momentum monoblock amplifier is hand-built and individually tested in D'Agostino's Connecticut factory. Rugged through-hole construction allows the use of higher-quality components, thus assuring superior performance for years to come.

Power	300W
Frequency Response	1Hz – 200kHz
Harmonic Distortion	0.03%
Signal-to-Noise Ratio	105dB
Gain	26.5dB
Power Consumption (standby)	<1W
Inputs / Outputs	XLR, RCA, High-quality binding posts
Options	Stands with spike legs
Dimensions (W x H x D)	12.5 x 4.3 x 18.5 inches / 318 x 109 x 470 mm
Weight	90lbs / 40.8kgs

Devialet

France
www.devialet.com

Optimised musicality, the best acoustic output, absolute harmonies. A Devialet has no fear of superlatives. By inventing the most faithful musical reproduction instrument in the world, they are constantly pursuing one clear target: transmit the 1 million nuances added by artists and composers, stored on a disc or digital file, to listeners with unerring neutrality, and intense and universal emotion. After 4 years of research and development, Devialet has achieved ADH®, a technology and audio system as revolutionary as high-definition for televisions. Expect to be astounded.

Devialet 240
Integrated Amplifier / DAC /
WiFi Streamer
€12,990

The 240, is the Devialet without compromise for enthusiasts. Its equipped with complete options, overpowerful, configurable as a 500W mono-block and totally upgradable. Its made with exclusive solid aluminum hand-polished frame. Devialet has also created the ideal remote your hand was designed to hold. Since the radio link is omnidirectional, wherever you sit your audio system is at your fingertips. They aimed to offer the right Devialet for you, and dreamt up three models, which will match your usages and your budget. But no compromise whatsoever was accepted for sound quality.

ADC	PCM4220
DAC	Burr-Brown PCM1792
Power	2 x 240W
Output Impedance	<0.001 ohms
Frequency Response	20Hz – 20kHz
Signal-to-Noise ratio	130dB
Harmonic Distortion	<0.001%
Intermodulation Distortion	0.001%
Dimensions (W x H x D)	15.7 x 1.6 x 15.7 inches / 400 x 41 x 400 mm
Weight	15.4lbs / 7kgs
Available Finishes	Dark Chrome, Silver

Eryk S Concept

Poland
www.eryksc.com

The Founder, Eryk, has spent twenty years seeking better design parameters for home audio systems, much of his research in this sphere concentrating on unique design concepts and system's configurations. He treats his designs as musical instruments, whose every string must be tuned to perfect harmony with others. With this in mind he concentrates on the object and sound transmission, the two hearts of every audio system. Eryk's audio designs have unique shapes based on good taste and ergonomics.

Red King
Integrated Tube-Amplifier
$2,690

The "Red King" is coming after the "Red Beauty" tube amplifier & this is serious deal: full remote control; three audio inputs including Bluetooth wireless connection with your mobile gadget; three big power supply stages for best sound quality & amp reliability; foremost this amp uses most hi-end mode in music reproduction: parallel single tube; arrangement to project spatial sound with full impact around your head, on all types of music. It took them almost two years to fine tune this concept.

Output Stage	EL84 tubes, PSE mode
Input Stage	NOS 6N6P, SRPP mode
Outputs	4 ohms & 8 ohms speaker terminals, RCA
Output Power	12W
Dimensions (W x H x D)	10.6 x 6.6 x 15.3 inches / 270 x 170 x 390 mm
Weight	30lbs / 14kgs
Available Finishes	Glossy Red

Erzetich Audio

Slovenia
www.erzetich-audio.com

Erzetich Audio is focused on personal audio, specifically on meticulously-hand-crafted high-quality headphone amplifiers and accessories. All of them are made in-house from scratch: from a raw plank of wood and bare PCB to a finished product. Erzetich never wanted their headphone amplifiers to be just electronic devices, but an object of art with a human touch - a tool to enhance the sonic part of your music enjoyment experience with an addition of human energy and dedication.

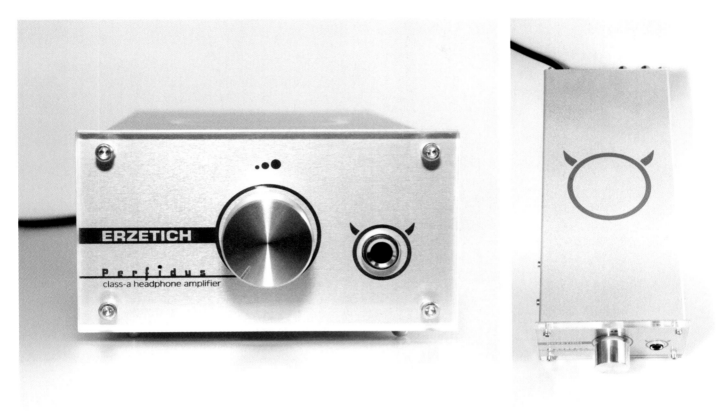

Perfidus

Headphone Amplifier
$1,240

It's fast! It's powerful! Its Perfidus! This class-A headphone amplifier will blow you away. Literally. It is so fast and powerful that you need to handle it with care. With its super detailed reproduction and speed, it will take your perception of music to another level. Classical orchestras will be more alive than ever and rock music will be beaty as it can be only on a live stage. Perfidus is your ultimate choice for your headset amplification.

Headphone Impedance	8 ohms - 600 ohms
Frequency Response	4Hz - 40kHz
Harmonic Distortion	0.007 (%)
Intermodulation Distortion	0.077 (%)
Power Consumption	15W
Power Supply	110V - 220V
Dimensions (W x H x D)	4.3 x 2.5 x 11 inches / 110 x 64 x 280 mm
Weight	3lbs / 1.33kgs

Fosgate

California, USA
www.musicalsurroundings.com

Jim Fosgate is the vision and expertise behind the Fosgate Signature. The manufacturers, Musical Surroundings is based in Oakland, California. It is also a distributor of high-end audio equipment. Since 1991, Musical Surroundings has specialized in turntables, tonearms, phono cartridges, phono stages, analog accessories and tube electronics for the high performance audio market. Countless man-hours of working with every possible major circuit approach was not just explored, but designed, built, and auditioned by Jim Fosgate himself.

Fosgate Signature Tube Amplifier

Headphone Amplifier
$1,500

The Fosgate Signature Tube Headphone Amplifier is designed by Jim Fosgate, a world-class audio engineer with over 18 audio-related patents, and manufactured for Musical Surroundings. Using 2 tubes and a special output buffer, to optimally match the wide range of headphone impedance, the Fosgate Signature can drive all of today's best headphones. It features Fosgate's unique and patented circuit designs, including Surround, that applies an out-of-phase cross-blend to create a sense of depth and space, moving the dimensional soundstage outside of your head, and Bass Boost to add extra weight and fullness to the lower octaves.

Frequency Response	2Hz – 200kHz
Voltage Gain	15dB
Signal-to-Noise Ratio	95dB
Impedance	30 ohms – 500 ohms
Maximum Output Voltage	26V
Maximum Power Output	180MW
Harmonic Distortion	0.05%
Maximum Bass Boost	9dB
Minimum Bass Boost	6dB
Dimensions (W x H x D)	7 x 5.5 x 11 inches / 178 x 140 x 279 mm
Weight	5lbs / 2.2kgs

Gato Audio

Denmark
www.gato-audio.com

Designed and built with beauty, simplicity, functionality and an extremely elegant user interface, Gato Audio electronics possess both brute strength and finesse to bring the subtlest musical details to life with precision and authority. Gato Audio loudspeakers are lavishly constructed of the finest materials with laminated cabinets, state of the art Danish high end drivers, and carefully calibrated heavy-duty crossover networks to provide breath-taking concert hall performance in a beautiful yet compact form.

DIA-250

Integrated Amplifier / DAC
€3,250

The Gato Audio DIA-250 Integrated Amplifier with digital inputs is designed and built to a vision of combining technology, power, sonic performance, connectivity, versatility, stunning looks and compact design. The DIA-250 amplifier offers a state of the art Class-D output stage with dedicated and optimized switch mode power supplies. The multiple stage power amplifier is built around proven technology from International Rectifier, optimized with a precision high frequency oscillator and PWM modulator.

Output Power	2 x 250W @ 8 ohms, 2 x 500W @ 4 ohms
Frequency Response	20Hz - 20kHz
Harmonic Distortion	<0.01%
Signal-to-Noise Ratio	<102dB
Gain	26dB
Recommended Load	4 ohms -16 ohms
Power Consumption (Standby)	<1 W
Connectors	Balanced/Unbalanced Analog I/Os ; Digital Inputs
Dimensions (W x H x D)	12.8 x 4.1 x 16.5 inches / 325 x 105 x 420 mm
Weight	22lbs / 10kg
Available Finishes	High Gloss – Black, Walnut, White

Hagerman

Hawaii, USA
www.hagtech.com

Hagerman Audio Labs is a small shop where they design and build heirloom quality audio products the old fashioned way. By hand. Sure, they use the latest technologies and processes, but never forget that extra level of care that can only be achieved through the human equation. All their designs are unique and innovative, pushing the state-of-the-art in audio reproduction. They are timeless and built to last.

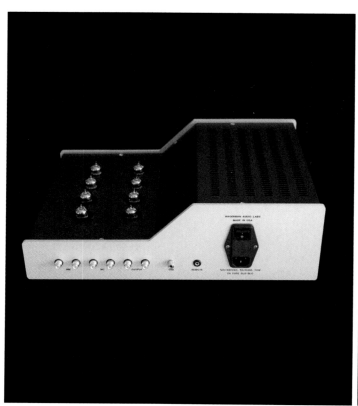

Trumpet Reference

Vacuum Tube Phonostage
$7,200

The Trumpet Reference is a truly reference-class phonostage for the most demanding of audiophiles. Based on the original Trumpet, it adds pushbutton control of all functions, a second input, and upgrades many of the components. The chassis has also been changed to reflect the reality of equipment racks by putting the power supply and amplifier boards in a side-by-side configuration. Status LEDs on the front panel indicate operational settings of the various functions. In addition to standard RIAA equalization, three other curves used in the LP "microgroove" era have been included. This combination of features, convenience, and sonic performance are unmatched at this price.

Tube Complement	4 x 12AX7, 4 x 12AU7
Gain	44dB (MM); 70dB (MC)
Input Impedance	47k ohms - 57k ohms
Output Impedance	800 ohms
RIAA Response	+/-0.5dB
Frequency Response	8Hz – 500kHz
Harmonic Distortion	<0.05%
Signal-to-Noise Ratio	70dB
Power	120VAC – 240VAC, 70W
Dimensions (W x H x D)	17 x 5 x 12 inches / 432 x 127 x 305 mm
Weight	19lbs / 8.6kgs

ISOL-8 Teknologies Ltd

United Kingdom
www.isol-8.co.uk

Nic Poulson founded Isol-8 Teknologies in 2003. The company has gradually gathered a steadfast following amongst those willing to open their ears, and they now produce a wide range of reliable, practical solutions, tailored to suit different applications across audio and AV. All the principal members of ISOL-8 are keen audiophiles. Together they have the balance of skills to innovate and deliver exceptional products. Above all they have the passion to push forward the standards of reproduced sound. This passion is what makes ISOL-8 the leading force in power conditioning.

ISOL-8 PowerStation

Twin Channel PowerStation
£2,499

The purpose of the PowerStation is simple: to get as close to the perfect low power supply as is practically possible. Two independent channels of clean, pure regenerated power will transform even the highest quality source components. At the heart of the PowerStation are three micro-controllers, which oversee operation and generate two ultra stable waveforms of purity that are simply unobtainable from your domestic supply. Twin power amplifiers provide output up to 100W per channel. They are specially designed for this role and are coupled to their load via bespoke output transformers.

Type of outlets	13A UK, Schuko, 15A US
Number of Outlets	2
Power Input	IEC Inlet
Mains Voltage	230VAC; 117VAC
Output Power	2 x 100W
Output Frequency	50Hz - 100Hz
Harmonic Distortion	0.05%
Power Consumption (Standby)	6W
Dimensions (W x H x D)	17.5 x 5.7 x 16.5 inches / 445 x 145 x 420 mm
Weight	41.8lbs / 19kgs
Available Finishes	Titanium

Kharma

Netherlands
www.kharma.com

Kharma embodies the seeking to ultimate beauty, where Kharma converges audiophile excellence with aesthetic beauty. A genuine approach to reveal the most subtle musical experience, and to unlock immeasurable aesthetic joyful inner experiences, in the listener. Like a piece of art in painting can evoke feelings of beauty and excitement in the viewer, the Kharma products are pieces of art in audio vibrations and they will evoke the highest feelings of beauty and wonder in the listener.

Exquisite Reference MP1000

Mono Power Amplifier
$48,000

New materials and new techniques gave Kharma the wings to further approach reality as the absolute reference in the Exquisite Collection. These new products have the same far-reaching attention to detail as the entire Exquisite Collection in both technological aspects and in the astonishing design. This mono power amplifier is a result of their unique reputation and style, which is realized by a team of highly dedicated and committed specialists. The Exquisite MP1000 amplifiers are state of the art solid-state amplifiers with a zero compromise approach.

Type	EXQ-MP1000-R
Circuit	Linear Amplifier Technology
Power	400W @ 8 ohms, 800W @ 4 ohms
Frequency Range	4Hz - 200kHz
Peak Power	2700W
Harmonic Distortion	-120dB
Inter-Modulation Distortion	-120dB
Connections	XLR / RCA
Nominal Impedance	10k ohms
Dimensions (WxHxD)	17.7 x 7.1 x 23.6 inches / 450 x 180 x 600 mm
Weight	165lbs / 75kg

KingRex Technology

Taiwan
www.kingrex.com

KingRex is a trend leading designer for digital stream devices, starting in the PC HiFi area. KingRex provides HiEnd cables as well. After they launched their first DAC with USB—T20U, they established a creative image in the audio industry. Along with providing high-end solutions, all of KingRex products are made to be high performance at a reasonable price. Their mission is to please your ear! Enrich your life! They are innovative by desire!

PreFerence

Phono Preamplifier
$1,350

PreFerence, the second-generation preamplifier from KingRex, is a logical extension of the success story of the award-wining Preamp. With KingRex featuring external power supply design; PreFerence is centered around high-quality op-amp and the diamond buffer, highly praised for its musicality. A current booster has been added to the final output of the amplifier stage to enhance drivability. DC servo circuit helps avoid unwanted audio signal loss due to coupling capacitors. Gain is user selectable with a choice of 6, 12 or 18dB. The result is accurate and perfect signal amplification, free from coloration and yet highly adaptable.

Input / Output	RCA
Gain	6dB, 12dB, 18dB
Frequency Response	20Hz – 20kHz
Signal-to-Noise Ratio	Better than 126dB
Harmonic Distortion	<0.0004%
Intermodulation Distortion	<0.0011%
Input Sensitivity	1.1 V
Power Consumption	<10W
Phonostage	MM, MC

KingRex Technology

Taiwan
www.kingrex.com

KingRex is a trend leading designer for digital stream devices, starting in the PC HiFi area. KingRex provides HiEnd cables as well. After they launched their first DAC with USB—T20U, they established a creative image in the audio industry. Along with providing high-end solutions, all of KingRex products are made to be high performance at a reasonable price. Their mission is to please your ear! Enrich your life! They are innovative by desire!

HQ-1

Headphone Amplifier
$850

HQ-1 uses the Class A MOSFET design. It is a general-purpose desktop headphone amplifier. Featuring a 2-box design, one is housing the power supply and the other enclosing the amplification circuitry. The output circuits are Toshiba A970/C2240 for first stage amplification and output stage by Hitachi MOSFET 2SK214. Power supply is a linear regulated type which has a total of 25,100uF of filtering capacitors. This ensures smooth DC power as well as a clean source for the sensitive amplification circuits.

Gain	18.5dB
Power Output	100mW @ 300 ohms, 170mW @ 150 ohms, 570mW @ 30 ohms
Output Impedance	8 ohms
Input Impedance	15k ohms
.Frequency Response	20Hz – 80kHz
Signal-to-Noise Ratio	115dB
Harmonic Distortion	0.005%
Channel Separation	>92dB
Dimensions	7.3 x 1.9 x 5.4 inches / 185 x 45 x 138 mm
Available Finishes	Red, Black

KR Audio

Czech Republic
www.kraudio.com

KR Audio Electronics is a manufacturer of hand-crafted vacuum tubes, custom designer tubes and high-end audio equipments. KR began making improved audio tubes after a 54 year break in the tube manufacturing sector. Any KR either as a tube up-grade or as an amplifier with the KR tube application using their hybrid circuit in the sound system, will give the listener more dynamics, greater detail from the recording be it CD, vinyl or audio streaming and an almost tangible imaging.

VA355i

Stereo Integrated Amplifier
$14,000

The VA355i continues to outperform the competition and impress the world over. This integrated amplifier delivers all the speed, resolution, dynamic energy and bass control of the best solid state amplifiers, but retains the tonal and spatial sweet sound of a world class tube amplifier. The VA355i lets you hear all of the transient details of the music. You can also hear the full harmonic bloom of the instruments and their very long decay. A truly natural perspective—they never run over each other and smear the sound. Very few amplifiers can walk this kind of fine line.

Design	Single-Ended Tube Amplifier
Amplification	Class A, Zero feedback
Output Tubes	2 x T100
Output Power	2 x 30W
Frequency Response	20Hz – 20kHz
Output Impedance	4 ohms, 8 ohms
Input	4 x 0.75V RMS
Input Impedance	47k ohms
Dimensions (W x H x D)	21 x 15.9 x 16.3 inches / 535 x 305 x 415 mm
Weight	81.5lbs / 37kgs

Krell

Connecticut, USA
www.krellonline.com

Founded in 1980, Krell Industries is the world's premier manufacturer of high performance audio and video equipment. Its amplifiers, preamplifiers, surround-sound processors, iPod docks and loudspeakers have won acclaim in countless reviews and among audiophiles worldwide. For decades, Krell Industries has been the world's best-known and most respected manufacturer of high-end audio products.

Foundation

7.1 Channel AV Processor
$6,500

Foundation, is the latest in a long line of state-of-the-art A/V processors from Krell. Featuring the latest digital connectivity, the Foundation also maintains Krell's exacting standards for analog audio reproduction. From decoding the latest lossless audio formats, extensive digital switching, 3DTV pass-through, and more, the Foundation is fully compliant with the digital age. Yet the 7.1-channel processor also features balanced audio outputs, automatic setup and room EQ, and Krell's legendary robust hardware. A slim new form factor exudes the Krell's aesthetic, while allowing placement in smaller equipment racks.

Signal-to-Noise Ratio	"A" weighted 106dB
Harmonic Distortion	-0.003dB
Maximum Input	6.3Vrms(Balanced) / 3.15Vrms(Single-ended)
Maximum Output	16.7Vrms(Balanced) / 8.35Vrms (Single-ended)
Input Impedance	58k ohms
Output Impedance	100 ohms
Power Consumption (Stand-by)	2W
Dimensions (W x H x D)	17.1 x 3.5 x 6.8 inches / 433 x 88 x 173 mm
Weight	18lbs / 8.16kgs

Lehmannaudio

Germany
www.lehmannaudio.com

Lehmannaudio is the German brand for high-end audio amplifiers for demanding music-lovers and professionals. In 1988, it all began with the passion for music. The company was founded by a young student of audio engineering, Norbert Lehmann, in Cologne, Germany. In everything that they do, they want to preserve the pristine sound of the musical piece. They have devoted themselves to the making of high-end audio devices which strive for perfection just as much as the artists whose music we enjoy.

Linear SE

Headphone Amplifier
$1,999

The next stage of the successful Linear headphone amplifier sets new benchmarks as to sound quality and the direct musical experience. The Linear SE lets you experience music over your headphone in a new way. Suddenly, never noticed details and the intention of the musicians are revealed. You will hear the music as a complete artwork and still plunge deep into the realm of individual sounds. Outwardly, the new Lehmannaudio champions league headphone amplifier stands out above all by a wealth of variants. The housing versions of the Linear SE range from satin mat aluminum with a distinguished aspect to high-class real wood veneers.

Frequency Response	10Hz – 35kHz
Signal-to-Noise Ratio	>95dB
Harmonic Distortion	<0.001%
Output Power	200mW
Power Consumption	10W
Input Impedance	47k ohms
Output Impedance	5 ohms; 60 ohms
Channel Separation	>70dB
Dimensions (W x H x D)	4.7 x 1.9 x 11.7 inches / 120 x 50 x 297mm
Weight	4lbs / 1.8kgs
Available Finishes	Black, Silver, Olive, Makassar, Zebrano, Wild Cherry, Oak, Walnut

LessLoss

Lithuania
www.lessloss.com

LessLoss is a high performance audio company. Since its internet inception five years ago, Less-Loss has shipped over 3000 high performance products to a worldwide customer base. LessLoss products have received press awards and over 200 reviews. Developed technologies include Skin-filtering for the reduction of noise, the use of Panzerholz to minimize micro-vibration, black body ambient field conditioning, as well as a breakthrough distortionless analog signal transfer technology called Tunnelbridge.

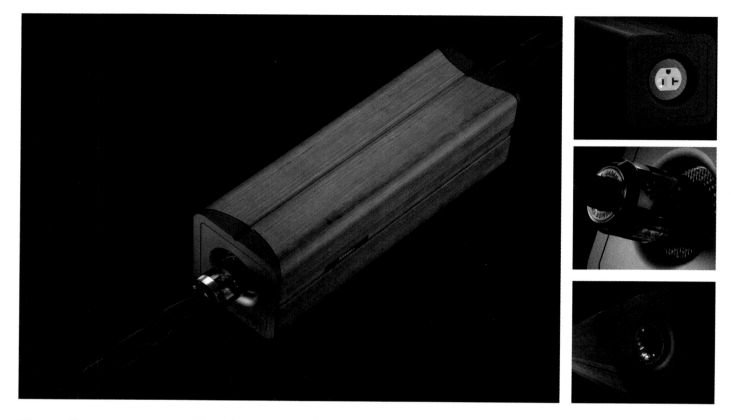

Firewall

Power Conditioner
$4,686

Featuring augmented LessLoss Skin-filtering technology, and a high performance panzer-holz acoustical damping structure, the Firewall blocks noise pollution to levels traditional capacitor and coil-based solutions can only hope to approach. Developed from LessLoss's critically acclaimed Dynamic Filtering Power Cable, the Firewall reveals hidden subtleties of the entire audio event as it eliminates the widespread negative effects of high frequency noise pollution.

Technology	LessLoss Flow Flux Skin-filtering
Outlet Options	USA/Japan type, European Schuko type
Input / Output	Single, Anti-vibration
Construction	Vibration Control: Panzerholz, Carbon Fiber, Anodized Aluminum Casing
Passive Elements	None
Options	DFPC cables
Dimensions (W x H x D)	4.3 x 4.4 x 13.8 inches / 110 x 113 x 35 mm

Linar Audio

Canada
www.linaraudio.com

Since its inception in 1998, Linar Audio is dedicated to the design and assembly of high performance audiophile amplifiers and pre-amplifiers. Linar products are the fruit of extensive Research and Development efforts in the audio field. Linar Audio products are maintaining the design philosophy with no overall feedback for stability and no capacitors in the signal path. If you are searching for the ultimate sound fidelity, then the Linar P107 should be on your short-list.

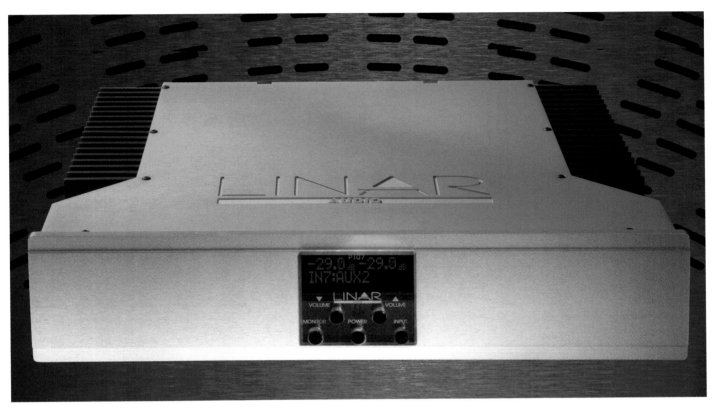

P107

Stereo Pre-Amplifier
$9,000

The Linar P107 Stereo Preamplifier offers the highest level of detailed sound possible. This is achieved by a straightforward design, free of signal path capacitors and the exclusive use of discrete transistors biased at their optimal operation point. of a passive volume control followed by a gain section without feedback. The volume control circuit uses mechanical relays with high precision low noise resistors to preserve the integrity of the signal. Continuing the tradition of Linar's designs, the P107 offers you flexibility in both configurations and connections. The detailed sound and flexibility of the Linar P107 make it the right choice for any high-end audio system.

Amplifier	Class A
Channel Separation	>100dB
Frequency Response	5Hz – 100kHz
Harmonic Distortion	<0.02%
Gain	12dB
Maximum Output Level	5Vrms
Signal-to-Noise Ratio	-103dB
Input Impedance	20k ohms
Output Impedance	50 ohms
Dimensions (W x H x D)	18 x 4.5 x 15.3 inches / 457 x 114 x 300 mm
Weight	28lbs / 12.7kgs

Lynx Studio

California, USA
www.lynxstudio.com

Formed in 1998 by a team of seasoned audio software and hardware engineers, their goal is to utilize cutting edge technology to create the highest quality products at a good value to the customer. The Lynx team is focused on linking the professional and high end home audio world with computers by utilizing their many years of experience in both hardware design and device driver coding. This complementary expertise is borne out in reliable products that not only have great specifications, but are also shipped with extremely stable and well-tested drivers.

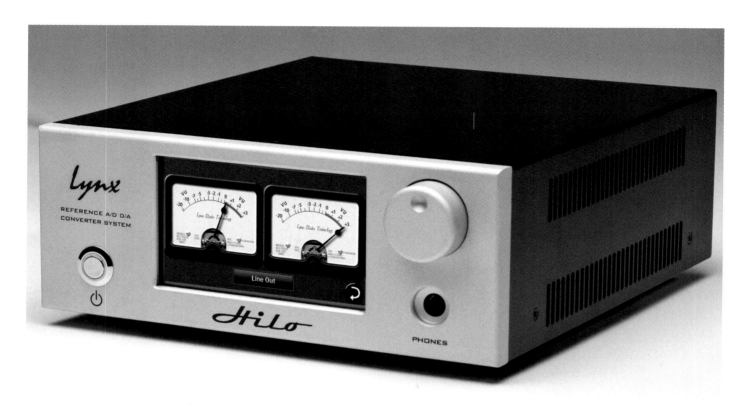

Hilo
Reference A/D D/A
Converter System
$2,495

Lynx Studio Technology is proud to present the Hilo Reference A/D D/A Converter System. With the pristine, open, transparent audio quality for which Lynx is known, Hilo provides two channels of mastering quality analog to digital conversion, up to six channels of digital to analog conversion, a secondary monitor output, and a world class independent headphone amplifier in a compact half-rack size. Initial front panel controls support extensive signal routing and mixing, sample rate selection, clock source options, levels, metering and diagnostic features.

A/D Harmonic Distortion	-114dB
A/D Crosstalk	-140dB
D/A Harmonic Distortion	-109dB
D/A Crosstalk	-135dB
Frequency Response	20Hz - 20kHz
Dynamic Range	121dB
Sample Rate	Up to 192kHz
Compatibility	Windows 32 bit, 64 bit / Macintosh OS X
LCD (High resolution/Touch Screen)	480 x 272 pixel
Dimensions (W x H x D)	8.5 x 3.3 x 10 inches / 216 x 83 x 254 mm
Weight	6.8lbs / 3kgs

McIntosh Labs

New York, USA
www.mcintoshlabs.com

Founded in 1949 and handcrafted in Binghamton, NY, McIntosh is a global leader in distinguished quality audio products. McIntosh continues to define the ultimate home entertainment experience for consumers around the world, with the iconic "McIntosh Blue" Watt Meters, recognized as a symbol of quality audio. Since its inception, McIntosh has been powering some of the most important moments in music history, including Woodstock and the Grateful Dead's legendary "Wall of Sound." McIntosh has not only witnessed history, it has shaped it.

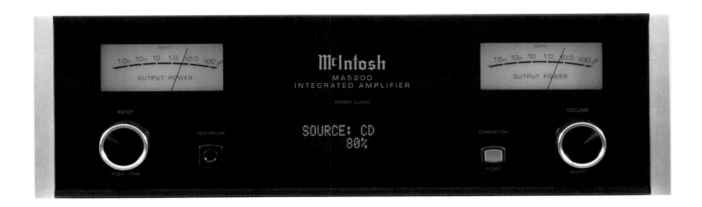

MA5200
Integrated Amplifier
$4,500

The MA5200 delivers the full McIntosh audio experience in a sleek, compact size. Its 100 watts per channel is perfect for modestly sized listening spaces. With 9 inputs, it provides you with enough connections and modern technology to begin building your home audio system. The Home Theater Pass Through feature will allow for seamless integration into your existing multi-channel theater system. MA5200 utilizes Power Guard® to prevent clipping and keep your speakers safe from damage. McIntosh's new High Drive headphone amplifier features increased gain and output power, and is optimized for virtually all headphone types for an enjoyable personal listening experience.

Frequency Response	20Hz – 20kHz
Output Power	2 x 100W
Harmonic Distortion	0.01%
Output Impedance	8 ohms
Signal-to-Noise Ratio	110dB
Input Impedance	20k ohms
Power Consumption (Standby)	<0.25W
Inputs	USB, Coaxial, Optical, Phono moving coil/magnet
Chassis	Dual layer
Dimensions (W x H x D)	17.5 x 6 x 22 inches / 445 x 152 x 559 mm
Weight	38lbs / 17.2kgs

Musical Surroundings

California, USA

www.a.musicalsurroundings.com

Musical Surroundings, based in Oakland, California, is both a distributor and manufacturer of high-end audio equipment. Since 1991, Musical Surroundings has specialized in turntables, tonearms, phono cartridges, phono stages, analog accessories and tube electronics for the high performance audio market. The companies that are associated with Musical Surroundings include Aesthetix, AMG Turntables, Benz Micro, Clearaudio, Fosgate Signature and Graham.

Nova II

Phono Pre-Amplifier

$1,200

The Nova II is a battery-powered phono stage with incredible flexibility and performance. The Nova II preamplifier features dual-mono, discrete circuitry with extensive gain and loading flexibility, delivering the best sound in its price range. It has rear panel accessible switches for easily accessible and highly flexible gain and loading adjustment. It features 2 rechargeable internal NiMH dual-mono battery packs with Smart Sensing auto recharge feature. All AC and charging circuits automatically disconnect when listening in battery mode.

Gain Switches	40dB - 60dB
Resistive Load	30 ohms - 100k ohms
Capacitive Load	200pF/300pF
Design	Discrete, dual-mono circuitry
Dimensions (W x H x D)	9.8 x 2.5 x 9.8 inches / 249 x 64 x 249 mm
Weight	5lbs / 2.2kgs
Available Finishes	Black, Silver

Nagra

Switzerland
www.nagraaudio.com

Nagra Audio designs, builds and markets a full range of portable digital audio recorders for Broadcast, Cinema, Music and Security professionals as well as a range of products for the high-end audiophile Hi-Fi industry. Audio Technology Switzerland (ATS), with headquarters in Romanel, Switzerland, was created in January 2012 and has subsidiary offices in Nashville, TN and Paris France. ATS designs, manufactures, and distributes the entire range of Nagra products for the high-end audiophile Hi-Fi industry.

Jazz
Pre-Amplifier
$12,250

The JAZZ is a Pure Class A vacuum tube preamplifier built-in a jewel-like machined anodized aluminum case with a solid aluminum front panel that offers input source selection, balance and volume controls, as well as a precision modulometer for system level matching. The JAZZ uses 3 high-grade vacuum tubes and Nagra-made transformers to provide irreproachable sonic clarity. The external power supply, ACPS II, is designed to deliver quiet, stable performance. The remote control boasts a fine tune balance control allowing minute differences in the stereo image to be delicately adjusted.

Frequency Response	10Hz – 50kHz
Harmonic Distortion	0.01%
Dynamic Range	>108dB
Crosstalk	>75dB
Inputs	XLR, RCA
Dimensions (W x H x D)	12.2 x 3 x 10 inches / 310 x 76 x 254 mm
Weight	7.2lbs / 3.3kgs
Available Finishes	Silver

NAT Audio

Serbia
www.nataudio.com

The main manufacturing program includes vacuum tube preamplifiers and power amplifiers. The main features of these products are triode configuration, short signal path, zero or low feedback and single ended concept. Special attention, in the process of fabrication, is in the choice of parts that are assembled in NAT audio components. All NAT audio components are hand made. NAT is one of the most innovative manufacturer globally in the segment of tube and hybrid high-end amplification.

Transmitter

Tube Monoblock Amplifier
$19,900

Transmitter is a new reference point power amplifier made by NAT standards – pure class A, single – ended type and zero global feedback circuit configuration. With power of over 100W in single – ended technique there is a capability to drive almost any type of loudspeaker. Hand crafted wide bandwidth output transformer (custom designed for NAT) is used in this unit. Output transformer contain no less then 61 separate different material layers. The quality of electronic material that is used inside Transmitter is impressive.

Type	Single Ended Class "A" Triode, Ultra-linear or Tetrode connection
Tube Complement	1 ' 6N1P-EV, 1 ' 6N30P/6N6P, 1 ' QB5/1750
Power Output	100W
Frequency Response	8Hz - 100kHz
Input Impedance	100k ohms
Input Sensitivity	1.5Vrms
Rise Time	3.0 microsecond
Noise	110dB below rated output "A" weighted
Power Supply	110VAC - 240VAC
Dimensions (W x H x D)	10.2 x 11.8 x 20.9 inches / 260 x 300 x 530 mm
Weight	77lbs / 40kgs

Norma Audio

Italy
www.normaudio.com

For more than 20 years Norma has been involved in the study and implementation of sophisticated audio amplifiers. Technical skill and musical sensitivity must be merged together, in the same way in which the best musical instruments are made. In Norma, we are convinced that the influence of audio electronics on the outcome of audio reproduction is much greater than commonly assumed. Listening to a NORMA product can express this concept better than words, closer than ever to live listening.

Revo IPA-140

Integrated Amplifier
$8,000

The Norma Revo IPA-140 integrated amplifier represents a definitive choice for amplification. Its high power and the ability to drive the most demanding speakers, make it an extremely versatile amp. Power, dynamics, speed and transparency are combined to create outstanding musicality and a total absence of listening fatigue. The elegant and sophisticated style is perfectly in tune with its sound personality, while the inside workings show that the technical refinement follows the same philosophy. The Revo IPA-140 uses a proprietary low noise circuitry, regulated high-speed power, and is built with a careful selection of the best materials available.

Electric Transformers	2 toroidal special audio use, 400 VA per channel
Inputs	4 RCA, 1 XLR Balanced, optional Phono, 1 USB DAC optional
Input Impedance	47k ohms
Output Impedance	200 ohms
Frequency Response	0Hz - 1.8MHz
Output Power	2 x 140W @ 8 ohms, 2 x 280W @ 4 ohms
Gain	34 dB
Dimensions (W x H x D)	17 x 4.3 x 14.3 inches / 430 x 110 x 365 mm
Weight	55lbs / 25kgs
Available Finishes	Silver & Black, Silver & Silver, Black & Black

Plitron Manufacturing

Canada
www.toruspower.com

Torus Power is precision engineered by Plitron Manufacturing Inc. Plitron has over 30 years of experience as an industry leader in toroidal transformer design and manufacturing. With its unparalleled reputation, Torus Power is manufactured under Plitron's ISO9001 medical-level quality control system in Toronto, Ontario. The genesis of Torus Power was provided by the combination of Plitron's capability to design and manufacture nocompromise toroidal transformers, with the deep understanding of the requirements of the audio and video industry for improved power quality to ensure performance and long-term reliability of all connected equipment.

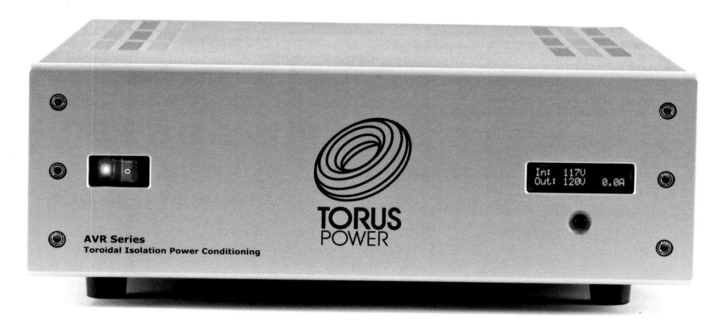

AVR 20

Automatic Voltage Regulator
$4,495

The AVR Series is the flagship line of Torus Power products, offering the full complement of features providing second-to-none performance and protection. The AVR 20 is rated for 20A of current output and isolates sensitive electronics from the surrounding electrical grid. Using proprietary Narrow Bandwidth Technology (NBT), the AVR 20 removes line noise. The AVR 20 also provides non-sacrificial surge protection from lightning strikes using Series Mode Surge Suppression (SMSS) and an Automatic Voltage Regulation and IP control to ensure that the connected system always receives a proper nominal voltage.

Output Current	20A
Number of Outlets	10
Rack Height	3U
Dimensions (W x H x D)	19 x 6.3 x 22.2 inches / 483 x 159 x 564 mm
Weight	101lbs / 46kgs
Available Finishes	Rack (19RK), Black (17CB), Silver (17CS)

Pureaudio

New Zealand
www.pureaudio.co.nz

How can we innovate in a way that breaks new ground? That was the question Gary Morrison and Ross Stevens asked themselves when creating Pureaudio. The result is a range of amplifiers that break the traditional mold by constantly asking 'is this the purest solution'. The distinctive exterior casing is produced from only two main parts. Circuit design uses the least number of parts and are all carefully chosen to serve the music. This synthesis of design produces striking looking products that deliver the highest order of musical satisfaction.

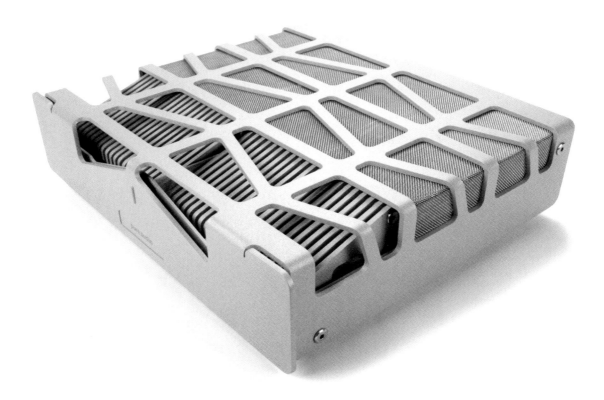

Reference Class A Monoblock

Monoblock Amplifier
$15,000

Pure musicality is what Pureaudio offer with their Reference Class A Monoblocks. Sounding neutral and devoid of any solid state signature, the amplifier is designed to work with domestic loudspeakers in typical home environments. The Class A Power rating is perfectly matched to this requirement and limits the energy consumption and heat generated during operation of a true Class A design. The power rating also allows the inclusion of a fully regulated, high-current power supply. Although operating in pure Class A, the amplifier falls back automatically to a low power standby mode when not in use.

Power	65W
Frequency Response	10Hz – 100kHz
Input Impedance	100k ohms
Harmonic Distortion	<0.01%
Gain	28dB
Power Consumption (standby)	5W
Current Output	30A peak
Control Terminals	12V
Hum & Noise	Inaudible
Dimensions (W x H x D)	18.9 x 4.5 x 16.1 inches / 480 x 115 x 410 mm
Weight	35.2lbs / 16kgs

Redgum Audio

Australia
www.redgumaudio.com

Specifically formulated to replicate the transients of music rather than just give big WRMS on test, REDGUM Audio's "amp with the key" grew into a range of Stereo and Home Theatre systems designed in-house by Ian Robinson. With the world's first audiophile CD Player with serviceable CDROM, the world's only Dual Mono Remote Control, only passive preamps, and fully analogue Home Theater processor, REDGUM's designs are unafraid to make some hard calls in the quest for sonic purity.

RGM Magnificata

Preamplifier/Power System
$28,000

After a 300W start in 1993 for their very first Special Order amplifier (RGM300), where could REDGUM go for a 20th Anniversary? Why not transform their biggest Signature Series amplifier (RGM300ENR pre/power system) into one purpose-designed to power ultra-low impedance speakers. Following a theme, there was no choice about its Eucalyptus name - it had to be the "Magnificata". Framed by three solid Red Gum wood fascias, attention to detail includes hand-wiring with single strand, solid core OCC Silver, 12 double-die solid state valves (Audiophile MOSFETs) per channel, all under the direction of its unique Dual Mono Remote Control.

Frequency Response	0.8Hz – 80kHz
Harmonic Distortion	0.009%
Intermodulation Distortion	<0.005%
Signal-to-Noise Ratio	>100dB
Input Impedance	10k ohms
Power Consumption (standby)	0.25W
Peak Current	>450W
Output Power	2 x 400W, 2 x 550W, 2 x 900W
Dimensions (W x H x D)	16.5 x 5.7 x 14.2 inches / 420 x 145 x 360 mm
Weight	112lb / 51kgs
Available Finishes	Solid Red Gum wood fascia, marine-grade EDP-coated / Steel Chassis

Roksan Audio

United Kingdom
www.roksan.co.uk

Founded in 1985, Roksan Audio has since developed into one of the best known and highly regarded names in audio. The company holds numerous worldwide industry awards and honors, making them continually one of the most successful companies in specialist high-level audio reproduction. Roksan's initial success was gained with the breakthrough Xerxes record player and progressed with a host of designs –from CD players and amplifiers through to streaming devices - that have become classics in the audio world.

Caspian M2

Integrated Amplifier
$2,610

The new M2 Integrated Amplifier advances the performance of this class of amplifier to a new level. The ease with which the M2 negotiates the loudspeaker load through even the most delicate yet demanding musical passages is almost unbelievable from such an elegant pure looking amplifier. Its control and authority appears to be backed by a state of the art power house! Resolution of detail is so harmonic and full of emotion that is reminiscent of the world acclaimed Roksan L1.5 reference preamplifier. Of course pedigree has played an important role but the relentless refinements, tests and never ending demands made this project come to life.

Input Impedance	47k ohms
Output Voltage	240mV, 700mV
Output Power	2 x 85W @ 8 ohms, 2 x 125W @ 4 ohms
Current Output	60A
Damping Factor	>160
Frequency Response	1Hz – 90kHz
Gain	40.2dB
Harmonic Distortion	0.002%
Power Consumption	<230W
Dimensions (W x H x D)	17 x 2.7 x 12.9 inches / 432 x 70 x 330 mm
Weight	28.6lbs / 13kgs

Roksan Audio

United Kingdom
www.roksan.co.uk

Founded in 1985, Roksan Audio has since developed into one of the best known and highly regarded names in audio. The company holds numerous worldwide industry awards and honors, making them continually one of the most successful companies in specialist high-level audio reproduction. Roksan's initial success was gained with the breakthrough Xerxes record player and progressed with a host of designs –from CD players and amplifiers through to streaming devices - that have become classics in the audio world.

Kandy K2

Stereo Power Amplifier
$695

The K2 stereo power amplifier is derived from the K2 integrated, and sets the new standard for affordable high-end power amplifiers. This power amplifier is refined way beyond expectation providing a powerful, punchy yet delicate and musical sound full of dynamics and timber. The sound stage and musicians have the realistic scale and 'air' about them that is achievable by only a few of the very best power amplifiers regardless of cost. The Kandy K2 range represents the starting point of Roksan Audio's portfolio and each component in the range delivers the highest possible audiophile quality musical reproduction available at the price.

Inputs	RCA
Input Impedance	23k ohms
Output Power	125W @ 8 ohms, 190W @ 4 ohms
Power Supply	500VA
Damping Factor	>110
Frequency Response	3Hz – 100kHz
Harmonic Distortion	<0.003%
Power Consumption	<25W
Signal-to-Noise Ratio	100dB
Dimensions (W x H x D)	17 x 3.5 x 14.9 inches / 432 x 90 x 380 mm
Weight	30.8lbs / 14kgs

Sanders Sound Systems

Colorado, USA
www.sanderssoundsystems.com

Sanders Sound Systems has been designing Electrostatic Loudspeakers (ESL's) and the amplifiers to drive them since 1974. Their reputation for honesty and dedication to customer service is due to their belief that satisfying customers is the best way to build a successful company. They offer a 30-day, in-home, risk-free trial. They will ship you any equipment that you wish. You can use it in your own home to listen to your familiar music, in your own listening room, with your own associated audio components for up to 30 days.

Magtech Amplifier

Amplifier
$5,500

The Magtech is the only amplifier with a linear, voltage regulator (patent pending). The Magtech "sounds as if it has infinite power into anything with total stability". The regulator in the Magtech amplifier maintains a stable voltage regardless of load or reasonable changes in the line voltage feeding the amplifier. Unlike other amplifiers, the distortion in the Magtech amplifier is virtually unchanged regardless of power level. The bias is stable regardless of load.

Power	2 x 500W @ 8 ohms, 2 x 900W @ 4ohms
Bandwidth	DC - 100kHz
Class of Operation	Class AB
Input Voltage	2V
Input Impedance	50k ohms
Gain	26dB
Harmonic Distortion	<0.004%
Intermodulation Distortion	<0.003%
Dimensions (W x H x D)	17 x 5.5 x 16 inches / 432 x 140 x 407 mm
Weight	55lbs / 25kgs
Available Finishes	Black, Silver

Soulution

Switzerland
www.soulution-audio.com

Purity of sound. No adulteration. No artificial enhancement. Nothing added. Nothing left out. It's a goal many audio manufacturers aspire to. But few attain. Soulution does. Without marketing gloss, just quietly understated design on the outside and cost-no-object technology inside, the result is audio for a truly discerning audience. Simply a uniquely sensual, organic experience that delivers all the honesty, the power, the subtlety and the emotion of music. Just as nature intended.

Soulution 530

Integrated Amplifier
$49,000

An uncompromising preamplifier and an uncompromising power amplifier, both of reference quality, combined in one housing. That is the way they at soulution understand the concept of an integrated amplifier. Enjoy the purity and power this music machine is able to deliver. It takes the uncompromising route of separate circuits, combining a dedicated reference quality preamplifier, and a dedicated reference quality power amplifier in one housing. It is a real Soulution product – truly a great amplifier.

Power	2 x 125W @ 8 ohms, 2 x 250W @ 4 ohms, 2 x 500W @ 2ohms
Power Consumption (standby)	<0.5W
Input Impedance	3k ohms
Frequency Response	DC - 800kHz
Gain	42dB
Harmonic Distortion	<0.001%
Signal-to-Noise Ratio	>120dB
Crosstalk	<110dB
Slew Rate	900nsec
Output Current	45A
Dimensions (W x H x D)	17.6 x 13.7 x 17.4 inches / 448 x 350 x 442 mm

Stahl~Tek

Texas, USA
www.stahltek.com

Stahl~Tek is a high-end electronics company founded in 2007. Stahl~Tek's driving force is an unrelenting passion for audio reproduction. That passion shines through in the products they design! The company was formed with the sole desire to design and build high-end audio components that immerse the listener in the music!

Ariaa
DAC
$12,900

Hailing from the heritage of the Opus Series, the Ariaa could have no better pedigree. Responding to the needs of industry, Stahl~Tek brought on additional audio engineers, who labored under the love of the craft to produce a work of art which matched both their desire for the purest audio as well as a more comfortable price point still within the reach of most audiophiles. The Ariaa is the result of their endeavor, a product of proven engineering.

Recommended Power	120/220VAC
Sampling rate	Up to 192kHz
Quantization	24-bit resolution
Frequency Response	0Hz - 100Hz
Signal-to-Noise Ratio	>127dB
Dynamic Range	>127dB
Harmonic Distortion	<0.005%
Output Level	3.25Vrms
Output Impedance	43 ohms
Dimensions (W x H x D)	17.6 x 15.7 x 4.2 inches / 447 x 400 x 106 mm
Weight	13lbs / 5.9kgs

Spiritual Audio

California, USA
www.spiritualaudio.net

At Spiritual Audio they strive to give you the audio/video quality that you deserve. With over 30 years of experience in the audio and electronic field, they only use the finest material and the best attention is put into each product. A new Spiritual Audio power station is the most important upgrade you can make in any audio system. Some say that your speakers are the weakest link in your audio system. But Spiritual Audio believe everything begins at the wall and state of the art power conditioning.

VX-12

Power Conditioner
$10,000

The VX-12 is for the extremist who demands the best sounding ac conditioner in the world. This high current 2500 watts/20- ampere device provides excellent performance, boasting 560 joules/30,000 amperes surge capacity with level two power factor correction. The VX-12 produces a huge 3-dimensional soundstage with thunderous tight bass and an extremely transparent open midrange. The top end is very detailed with tons of air around the vocals and instruments. The VX-12 transforms listening to music into a more defined and genuine experience, giving it an unparalleled organic sound.

Input Power	125VAC
Amp Fuses	2 x 10 Amp Fuses
Power Consumption	3W
Dimensions (W x H x D)	19 x 5.3 x 9.2 inches / 483 x 135 x 234 mm
Weight	20lbs / 9kgs
Available Finishes	Silver

Sutra

Italy
www.sutra.it

Sutra was born out of passion. Conceived and designed first of all for us, made and assembled by hand, piece by piece, each Sutra is like a small creature, the product of a merger between the deep commitment to research in the field of sound and love for music. The circle, party with friends and acquaintances, was thus extended to all those who have granted the time to listen, all who now have a Sutra to fill the spaces of the house with their favorite music.

Sutra 1.3

Integrated Amplifer
$1,300

Sutra 1.3 is an integrated amplifier equipped with four selectable inputs. Designed by engineers at Sutra, the product is the result of in-depth research of an architecture that could maintain the highest levels of efficiency and low distortion, while minimizing cost and size. Class D technology was the winning choice for small 1.3. Through the use of finely selected components and the search for solutions advanced engineering, Sutra will give the purity of sound you want, giving you an ever-more balanced and exciting experience combined with the convenience and reliability that only an amplifier of this type can provide.

Frequency Response	20Hz – 20kHz
Harmonic Distortion	0.04%
Dynamic Range	98dB
Output Power	2 x 15W @ 4 ohms, 2 x 10W @ 8 ohms
Power Consumption	35W
Toroidal Transformer	60VA
Bulk Capacitance	>30,000 MicroFarad
Analog Inputs	RCA, 13.5mm jack
Dimensions (W x H x D)	9.4 x 3.5 x 8.4 inches / 240 x 90 x 215 mm
Weight	6.4lbs / 2.9kgs
Available Finishes	Black, White

Synthesis

Italy
www.synthesis.co.it

Synthesis is a dynamic, fast growing company that looks to the future and, compared with other companies in this field, has been able to anticipate the metamorphosis of the various success factors in the Hi-Fi market. From its foundation in 1992, the "Synthesis Art in Music ®" has had as its main objective, the design of equipment whose ultimate aim is not just sound reproduction, but also to offer a product that represents refined elegance and musicality all rolled into a single element.

Action A100T
Integrated Stereo
Tube Amplifier
$7,600

The A100T is an integrated stereo tube amplifier with DAC on board, guaranteeing exceptional sonic performance from any digital source. Made with the finest materials & advanced technology, this series is a sophisticated blend of research & experience, designed without compromise for a rich sound & faithful music reproduction. The A100T provides a minimum of 200 watts power output from 8, KT66's in Ultra-Linear (UL) - parallel push/pull operation. The A100T features a newly designed custom output transformer resulting in a rich sound stage but with an authority that belies its 100Wpc.

Power Stage	4 x KT66
Input Stage / Driver	12AX7, ECC83 / 12BH7
Frequency Response	20Hz – 20kHz
Signal-to-Noise Ratio	>90dB
Output Power	2 x 100W
Inputs	CDP, DVD, Radio, SAT, USB, DGT
Input Impedance	50k ohms
Output Impedance	6 ohms
Dimensions (W x H x D)	16.5 x 10.2 x 17.7 inches / 420 x 260 x 450 mm
Weight	88lbs / 40kg
Available Finishes	Steel

Synthesis

Italy
www.synthesis.co.it

Synthesis is a dynamic, fast growing company that looks to the future and, compared with other companies in this field, has been able to anticipate the metamorphosis of the various success factors in the Hi-Fi market. From its foundation in 1992, the "Synthesis Art in Music ®" has had as its main objective, the design of equipment whose ultimate aim is not just sound reproduction, but also to offer a product that represents refined elegance and musicality all rolled into a single element.

Brio

Moving Magnet Phono Stage
$1,500

Brio phono stage caters for both MM and mid-high level MC cartridges with high output impedance. Separate power supply reduce the interferences from a power line. All of the units are designed in order to be able to reproduce music in normal domestic atmospheres, rather than only in an acoustically perfect room. Great care has been given to the aesthetic factor. Beautiful real woods and matched finishes make these units a perfect partner with the other fine products of Synthesis. With such a choice this allows, with both great elegance and refinement, these systems to integrate seamlessly into any home environment.

Input / Output Stage	½ 12AX7 - ECC83
Input Impedance	47k ohms
Gain	40dB
Frequency Response	20Hz – 20kHz
Signal-to-Noise Ratio	>80dB
Phase	Non-Inverted
Power Consumption	15W
Maximum Output Level	10V RMS
Dimensions (W x H x D)	5.7 x 3.9 x 8.6 inches / 145 x 100 x 220 mm
Weight	3.3lbs / 1.5kgs
Available Finishes	Ecologically sensitive lacquer

Synthesis

Italy
www.synthesis.co.it

Synthesis is a dynamic, fast growing company that looks to the future and, compared with other companies in this field, has been able to anticipate the metamorphosis of the various success factors in the Hi-Fi market. From its foundation in 1992, the "Synthesis Art in Music ®" has had as its main objective, the design of equipment whose ultimate aim is not just sound reproduction, but also to offer a product that represents refined elegance and musicality all rolled into a single element.

Matrix

Tube DAC
$3,500

The Matrix is designed completely in house by Synthesis to represent a new level in audio realism. It represents three years of meticulous work from drawing board to finished item. Why have they taken so long? Because they didn't want to create a DAC that boasted the worlds most impressive features, or heralded a new level in stereo detail, they didn't even want it to look like it just landed from Saturn with swages of aluminum and more lights than an oil rig at night. What Synthesis wanted, was to get back to the very basics of music itself.

DAC	2 x Wolfson WM8740
Output Stage	Philbrick Op-Amp K2-W 4 x 12AX7/ECC83
Frequency Response	20Hz – 20kHz
Signal-to-Noise Ratio	>95dB
Upsampling	192kHz / 24 Bit
Inputs	RCA, Optical Toslink, USB
Compatible Formats	PCM Stereo 32 – 192kHz
Power Consumption (Standby)	4W
Dimensions (W x H x D)	12.5 x 2.3 x 8.6 inches / 320 x 60 x 220 mm
Weight	11lbs / 5kgs
Available Finishes	Ecologically sensitive lacquer

Tangent Audio

Denmark
www.tangent-audio.com

Tangent was established in the 70's by a group of dedicated hi-fi enthusiasts. Within few years, innovative designs established the company as a major specialist loudspeaker manufacturer. Consistently, its products were highly recommended by respected hi-fi reviewers both from the United Kingdom and beyond. In 2005 the Tangent brand was re-introduced, now also including hi-fi separates and table radios. Over the years, Tangent has shown that fundamentally correct design can produce a product capable of achieving excellent performance at an affordable price.

EXEO
Amplifier
€299.95

Stop and take a look around. No matter how you slice it, most of us should agree that life is good. But can you make a good thing better? Sure, because even the improvements that seem small add up over time in a positive way. Take the new Tangent EXEO, for example. It consists of only two audio components, yet provides a number of connection and playback possibilities that indeed all add up to a lifetime of positive experiences. Centrally located, the VFD display offers a wide viewing angle and an appealing intensity. Highlighting user-friendliness, the display can be dimmed according to the ambient light.

Settings	Bass, Treble
Outputs	Record out, Sub-out, Pre-out
Inputs	Aux 1, Aux 2, CD, Phono MM, NET, DAB
Frequency Response	20Hz – 20kHz
Power	2 x 100W @ 4 ohms, 2 x 60W @ 8 ohms
Signal-to-Noise Ratio	101dB
Cross talk	70
Voltage	220VAC
Dimensions	16.9 x 3.6 x 11.1 inches / 430 x 92 x 283 mm
Weight	18.7lbs / 8.5kgs
Available Finishes	Aluminum Black

TBI Audio Systems

United Kingdom
www.tbisound.com

The Company's mission is to provide high value, high quality, and natural sound audiophile products. Their audio monitors and subwoofers are supported by patented cutting edge technology focusing on some of the more complex issues associated with subwoofers and loudspeakers. Patented ETL technology allows for breakthroughs in room placement options, soundstaging and off axis imaging with outstanding rendition of detail. You and your friends get an excellent listening position. Browse through their site for more information on TBI's exciting product lineup.

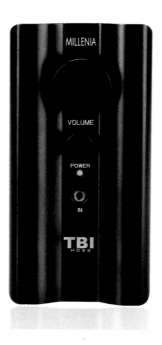 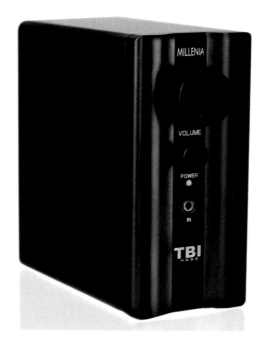 

Millenia MG3

Integrated Amplifier
$500

Introducing the exciting new Millenia MG3 from TBI Audio Systems where quality, audio performance, execution and mobility take front stage. This amplifier represents the state of the art in its construction and execution of linear and class BD electronic technology for a new level of audio enjoyment at home and on the go. While this new amplifier was designed to complement TBI's HDSS® based Majestic Diamond Audio Monitors it will benefit any quality speaker system. The Millenia MG3 is green by design seeking out all energy saving measures as an integral part of its operation.

Class of Operation	Class BD
Input Impedance	75k ohms
Output Impedance	8 ohms
Harmonic Distortion	<1%
Output Power	2 x 8W @ 8 ohms, 2 x 32W @ 8 ohms
Dimensions (W x H x D)	2.7 x 5.7 x 6.5 inches / 70 x 148 x 165 mm
Weight	2.3lbs / 1kg

TeddyPardo

Israel
www.teddypardo.com

The goal set by TeddyPardo is to provide a pure music listening experience that evokes the deepest emotions, and brings you closer than ever to the music. To assure the highest possible quality they developed proprietary measurement and test equipment, and Teddy Pardo personally continue to test each and every product before it is being shipped. TeddyPardo develop high end audio products that deliver an accurate, realistic and natural sound, with a sound signature that appeals to the most demanding audiophiles.

TeddyDAC-VC

DAC

$1,449

The TeddyDAC reproduces the most accurate, realistic and natural sound, free of digital artifacts, and achieves an "analog" sound signature that appeals even to the most demanding LP and vacuum tube amplifier owners. To achieve this goal, a unique approach was taken based on the understanding that the power supply is a major, if not the most critical, ingredient of the design. With the TeddyDAC, human voices sound as they sound in real life, orchestral music sounds detailed and three dimensional, just like in concert halls. The TeddyDAC contributes to many enjoyable listening hours without fatigue.

Inputs	2 x Coax, optical, USB
Voltage	110V - 120V / 220V - 240V
Sample Rates	Up to 192kHz
Output Levels	2V (RCA)
Options	192khz Async USB, Fully balanced XLR, Remote Control
Dimensions (W x H x D)	6.7 x 2.4 x 9.8 inches / 170 x 61 x 249 mm
Weight	4.8lbs / 2.2kgs
Available Finishes	Black

TeddyPardo

Israel
www.teddypardo.com

The goal set by TeddyPardo is to provide a pure music listening experience that evokes the deepest emotions, and brings you closer than ever to the music. To assure the highest possible quality they developed proprietary measurement and test equipment, and Teddy Pardo personally continue to test each and every product before it is being shipped. TeddyPardo develop high end audio products that deliver an accurate, realistic and natural sound, with a sound signature that appeals to the most demanding audiophiles.

TeddyPre PR1
Preamplifier
$1,749

The PR1 is designed around the concept that the audio circuits should be minimalistic with the shortest possible signal path, while the complexity is moved to the power supply. Although the PR1 is a solid state pre amplifier, its design is based on concept taken from valve amplifiers and its sound characteristics resemble those of high end valve amplifiers, but without the inconvenience of heat and wear. The PR1 is built using the best available components: selected low noise JFET transistors, low noise Tantalum, PPS capacitors, and WBT NextGen sockets.

Inputs	4 x aux, 1 x home theatre bypass
Outputs	1 x RCA, Second output optional
Voltage	110-120V / 220-240V
Gain	x6
Remote Control	Volume Control
Options	Power Supply included in a separate box
Dimensions (W x H x D)	6.7 x 2.4 x 9.8 inches / 170 x 61 x 249 mm
Available Finishes	Black

Trinity
Electronic
Design

Germany
www.trinity-ed.de

Trinity's endeavor to bring musical masterpieces from concert hall to living room without loss of quality is the engine that drives Trinity's development. Decades of experience in the field of electronics brought them to the point where they were able to manufacture the technically unrivaled high-end devices for which the name Trinity stands. They concentrate on design that perfectly implements what is practical and simple perfection.

Phono Stage

Pre-Amplifier
€24,900

Simplicity is the essence of brilliance. 96 low noise Integrated Circuits connected in a new and unprecedented circuit architecture ensure an absolute distortion free playback of your LPs. This new design works with all MC pickups and does not need, even at low input levels of 150µV, a sound-impairing input transformer. The noise of the Phono is well below the system related noise of an empty groove.

Inputs	2 x XLR Inputs
Equalizer Curves	RIAA, FFRR, EMI
Frequency Response	20Hz - 20kHz
Gain	72dB
Harmonic Distortion	Distortion free
Output Polarity	Non-Inverting
MC-Load Connector	5, 10, 25, 50, 75, 100, 200, 250, 300, 500
Outputs	XLR
Power Supply	100VAC - 240VAC
Dimensions (W x H x D)	17.3 x 2.5 x 11.8 inches / 440 x 65 x 300 mm
Weight	20lbs / 9kgs

Trinity
Electronic
Design

Germany
www.trinity-ed.de

Trinity's endeavor to bring musical masterpieces from concert hall to living room without loss of quality is the engine that drives Trinity's development. Decades of experience in the field of electronics brought them to the point where they were able to manufacture the technically unrivaled high-end devices for which the name Trinity stands. They concentrate on design that perfectly implements what is practical and simple perfection.

PREAMP
Pre-Amplifier
€24,900

The excellent circuit architecture with 48 low noise Integrated Circuits similar to the Trinity Phono Pre-Amplifier is also used in the Trinity Pre-Amplifier. Combined with the fully-balanced discrete volume control, it creates an innovating concept for sophisticated audiophile requirements. The PREAMP sets new standards to allow discerning music experts to enjoy at last their best recordings in true-to-life sound.

Inputs	4 XLR Inputs/channel
Outputs	2 XLR/Channel
Frequency Response	20Hz - 100kHz
Harmonic Distortion	>0.0001%
Gain	6dB
Discrete Volume Control	64dB in 1dB steps
Signal-to-Noise Ratio	-119dB
Output Polarity	Non-Inverting
Power Supply	100VAC - 240VAC
Dimensions (W x H x D)	17.3 x 11.8 x 2.5 inches / 440 x 300 x 65 mm
Weight	20lbs / 9kgs

Viola Audio Laboratories

Connecticut, USA
www.violalabs.com

In 2001 Paul Jayson and Tom Colangelo founded Viola Audio Laboratories to further their quest for accuracy and sonic neutrality. Viola, at the pinnacle of audio excellence, has advanced audio reproduction to an even higher standard. Years in development, Paul Jayson has brought together the most innovative digital technology and the highest quality, hand selected parts and components assembled in a uni-body all aluminum package. The look of refined power. You should hear how it makes you feel.

Crescendo
Pre-Amplifier
$22,500

With the introduction of the Crescendo, Viola Audio Laboratories has again elevated what the true audiophile comes to expect from their audio components. The Viola Crescendo preamplifier is a single chassis stereo preamplifier. The Crescendo chassis is milled from a solid billet of aircraft grade Aluminum, resulting in a very rigid chassis with very low resonance. The Crescendo features a high resolution USB DAC input. The Crescendo uses Wi-Fi remote control on Apple devices.

Outputs	1 RCA, 1 XLR, 1 Fixed (Tape)
Output Impedance	Main output 100 ohms / Tape output 600 ohms
Gain	16dB, 26dB switchable
Maximum Output	14.6V rms balanced / 7.3V rms single-ended
Frequency Response	20Hz - 20kHz
Intermodulation Distortion	<0.005%
Harmonic Distortion	<0.01%
Noise	>-90dBv
Power Consumption	37W
Dimensions (W x H x D)	17.5 x 3.5 x 15 inches / 445 x 89 x 381 mm
Weight	25lbs / 11.3kg

Wadia

New York, USA
www.wadia.com

For over two decades now, Wadia has focused on one goal: developing and applying technology to provide the most engaging and realistic musical performance from digital sources. Their philosophy is straightforward - they believe there is musical magic encoded in the numbers stored on your discs and drives. Wadia has spent years compiling an array of concepts and techniques to make their products extract, decode, and reconstruct the subtle cues to transform sound into music.

Intuition 01

Integrated Amplifier
$7,500

Wadia has advanced its tradition of progressive technology and industrial design to deliver an integrated amplifier with the utmost performance and aesthetics. With 9 inputs, including a USB input that supports even DSD playback, the Intuition 01 will connect you to all of your music. A powerful processor that up-samples incoming data, and an output stage capable of producing up to 350 watts of power, enables the Intuition 01 to drive virtually any loudspeaker with an immediacy of sound as you have never heard before.

DAC	ESS 9018
Frequency Response	3Hz – 45kHz
Signal-to-Noise Ratio	113dB
Jitter	<1ps
Harmonic Distortion	<0.005%
Inputs	USB, Optical, Wadialink, AES/EBU, Coaxial, RCA
Power	2 x 350W @ 4 ohms , 2 x 190W @ 8 ohms
Upsampling	1.536MHz
Dimensions (W x H x D)	15 x 2.3 x 15 inches / 380 x 60 x 380 mm
Weight	13.2lbs / 6kgs
Available Finishes	Black, Silver

Wadia

New York, USA
www.wadia.com

For over two decades now, Wadia has focused on one goal: developing and applying technology to provide the most engaging and realistic musical performance from digital sources. Their philosophy is straightforward - they believe there is musical magic encoded in the numbers stored on your discs and drives. Wadia has spent years compiling an array of concepts and techniques to make their products extract, decode, and reconstruct the subtle cues to transform sound into music.

Wadia 121

DAC / Pre-Amplifier
$1,299

The Wadia 121 decoding computer is a high performance DAC/digital preamplifier with a complete set of digital inputs, balanced analog outputs, digital volume control, and a headphone output section. Just connect your iTransport, computer, streaming audio source, or any other digital source and experience exceptional audiophile quality sound. All inputs including USB accept up to 24-bit/192kHz input data rates. The Wadia 121 decoding computer provides true balanced and single ended analog outputs. An exceptional headphone amplifier section delivers an unsurpassed personal listening experience.

Digital Inputs	XLR, RCA, BNC, Toslink Optical, USB
Input Data Rates	Up to 192kHz
Analog Outputs	Balanced XLR, Unbalanced RCA
USB System Requirements	Apple OS10.6.4, Windows XP SP3, Windows Vista SP2, Windows 7
Dimensions (W x H x D)	8 x 2.7 x 8 inches / 20 x 7 x 20 mm
Weight	2.5lbs / 1.1kgs
Available Finishes	Black

Wyred 4 Sound

California, USA
www.wyred4sound.com

Merging in-house design with manufacturing, Wyred 4 Sound produces hi-fi gear ranging from reference DAC's to kilowatt mono-blocks. They pride in the ability to offer these products at an incredible value with unparalleled performance. Having engineering strengths to this caliber is an added benefit to their unequaled support. Exceeding their customer's expectations is how they build relationships that last for years to come. Take a moment to find out how you can join the W4S family.

mPre & mAmp

Pre-Amplifier and Amplifier
$1,099 + $899

Introducing a DAC/Pre/Amp combo that is as surprisingly compact as it is powerful. The mPRE and mAMP are sculpted after principles taken from their highly acclaimed components and emphasized on key elements that giant killers are born from. Discrete audio circuits and highly efficient class D power amps bring your system to the cusp of audio technology. Already making waves around the world, these components offer a tonal balance with a touch of tube warmth and the speed/power of a modern day product. With many products to choose from, W4S is what you are looking for!

Frequency Response	20Hz - 20kHz
Gain (mPre)	17.2dB
Gain (mAmp)	30.5dB
Output Power (mAmp)	255W @ 8 ohms, 430W @ 4 ohms
Signal-to-Noise Ratio (mPre)	>105dB
Harmonic Distortion (mPre)	<0.006%
Harmonic Distortion (mAmp)	<0.002%
Output Impedance (mPre)	100 ohms
Output Impedance (mAmp)	0.018 ohms
Dimensions (W x H x D)	8 x 3.5 x 8 inches / 203 x 89 x 203 mm
Weight	8lbs / 3.6kgs

Zesto Audio

California, USA
www.zestoaudio.com

George Counnas, President of Zesto Audio loved music all his life – as a musician, record producer and audio engineer. That makes him fussy about sound. He started doing research on past tube designs going back to the original RCA circuits of the 1930's. Their "classic" designs were his starting point. To make sure its visual aesthetics lived up to its sound, he had an industrial designer and an artist collaborate to create an elegant enclosure. In the end, however, it is about what you hear because hearing is believing.

Bia 120

Stereo Power Amplifier
$12,500

What is Bia? - In Greek mythology, Bia is the Titan Goddess of force and power. With pro background in the recording studio and live concert mixing, Zesto Audio wanted to build an amp that allowed them to hear music the way they know it should sound. In short, they wanted power without compromising the music. The tubes of the Bia 120 amp are always on, so you have power when you need it. Class A is fast and responsive. There is no lag, providing your music more punch, transient response, better dynamic range and transparency.

Class of Operation	Class A
Design	Push-Pull, Ultra-Linear
Output Power	2 x 60W
Output Impedance	4, 8, 16 ohms
Polarity	Non-Inverting
Inputs	XLR, SE, BAL, 2V RMS inputs
Frequency Response	20Hz – 50kHz
Harmonic Distortion	0.22%
Gain	23dB
Dimensions (W x H x D)	17 x 10 x 20 inches / 432 x 254 x 508 mm
Weight	66lbs / 30kgs

TURNTABLES | MUSIC PLAYERS

128
Acoustic Solid
Germany

129
Althea Musica
Switzerland

130
AMG
Germany

131
Arcam
United Kingdom

132
Arcam
United Kingdom

133
Astell&Kern
Korea

134
Audiowood
Florida, USA

135
Baetis Audio
Montana, USA

136
Burmester
Germany

137
CasaTunes
Florida, USA

138
Canor Audio
Slovakia

139
Clearaudio
Germany

140
Hoerning Hybrid
Denmark

141
McIntosh Labs
New York, USA

142
McIntosh Labs
New York, USA

143
Oracle Audio
Canada

144
Origin Live
United Kingdom

145
Pro-Ject Audio
Systems
Austria

146
PTP Audio
Amsterdam

147
Roksan Audio
United Kingdom

148
Rotel
Japan

149
Soulution
Switzerland

150
Symbol
New York, USA

151
Synthesis
Italy

Acoustic Solid

Germany
www.acoustic-solid.com

Acoustic Solid was founded in 1997. Under this label, record players, luxury record players and tonearms in the high-end turntable segment are developed by hand in a real manufacturing plant, then manufactured and marketed worldwide. Acoustic Solid produces unique, world-class products at the highest level of excellence. Take their Machine Small turntable, for example. Its shape and design – both visually and acoustically – has set an absolute standard for turntable drives. Their philosophy: Music is like family: harmony comes first. And music is life!

Solid Royal

Turntable
$17,000

The Acoustic Solid Royal is the company's state of the art model and represents the culmination of their development to date. A separately located motor isolates the platter from any motor induced vibration. Provision for three tone arm bases is provided to allow the maximum flexibility in system configuration. The Acoustic Solid Royal provides analog playback of great naturalness and musicality. Stereo images are focussed and stable; dynamics are impressively realistic and background noise is negligible. When used in a system of the appropriate quality, you really can feel that the musicians are in the room and playing for you.

Turntable Platter	70mm Aluminum
Tonearm	WTB211 arm
Cartridge	Ortofon SPU Royal N Moving Coil
Motor	24V Berger Synchronic motor with M4 and PSU
Drive	String-drive in separate housing
Mat	Genuine Leather Royal blue, 5mm acrylic layered plate
Base	80mm thick aluminum body, Can accept up to 3 tonearms
Options	Tonearms, Cartridges and Stands as per your choice
Dimensions	17.3 x 9.5 x 17.3 inches / 440 x 241 x 440 mm
Weight	110lbs / 49.8kgs
Available Finishes	Polished, Satin

Althea Musica

Switzerland
www.altheamusica.com

Althea Musica stands for experience of musical emotions in its highest form. The presented performance of an artist and his dialogue with the composition of the work opens a new dimension of the sound stage in all its colors, strength, warmth and sonority. Every single component of the system embodies the basic values of the manufacture and the soul of her musical structure is recognizable on the first auditioning. A successful symbiosis of elegant design, sophisticated technology and music pleasure.

Vega

Turntable
€48,000

A jewel of the analogous turntable construction – the nicest way to listen to your beloved disks! The form and function, the being and its appearance form a perfect harmony in the highest completion. It's nobly and timeless in the design, as well as enticing in the sound - a successful synthesis from mechanical skills and engineer's ingenuity. With its geometry and material choice, every swinging inclination is virtually suppressed - a masterly achievement of fine-mechanical precision from 40-year-old experience by bearing design. Because of its separate placement, any tone arm with length and geometry can be mounted.

Plinth	Combination of tone woods; Multiplex-flat construct
Platter	ALCOA – Aluminum Alloy
Bearing	Magnetic Bearing - Made of bearing bronze and hardened steel
Speed	33.3rpm, 45rpm
Tonearm Length	Any length
Power Supply	DC for Motor controller
Plinth Dimensions (Diameter x Height)	11.8 x 8.2 inches / 300 x 210mm
Motor & Tonearm Stands (D x H)	5.1 x 5.9 inches / 130 x 150mm
Weight	66lbs / 30kgs
Available Finishes	Black

AMG

Germany
www.amg-turntables.com

Werner Roeschlau, his son, Julian Lorenzi and other master machinists developed the AMG (Analog Manufaktur Germany) products at their bespoke multi-story Bavarian factory located north of Munich. Their factory has been manufacturing key, precision parts for some of the world's most highly regarded turntables for over a decade. This expertise in the design and manufacture of turntables led to the AMG line, premiering with the Viella 12.

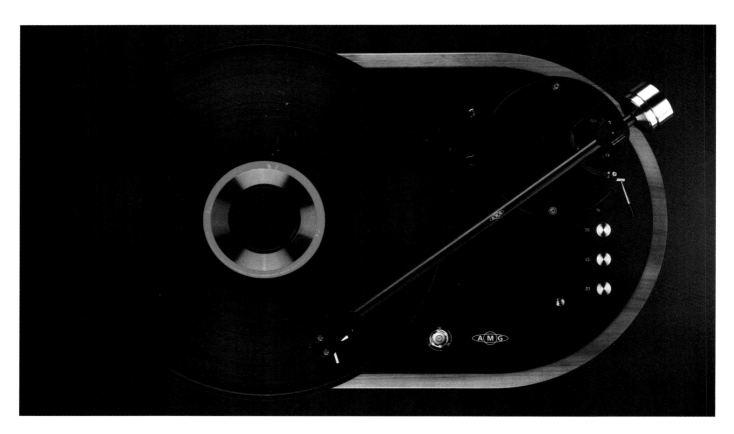

AMG Viella 12

Turntable
$16,500

Precision engineering and classic design are embodied in the first turntable from AMG, the Viella 12 or simply, V12. The AMG turntable line was created by a group of audio industry experts to advance the art of vinyl playback. All machining is done in-house, combining the latest Computer Aided Design and CNC machines with "classic analog" tools, including custom lathes and drill presses.

Type	Unsuspended, belt-driven turntable
Plinth	CNC machined in house using 25mm aircraft grade Aluminum
Platter	Decoupled Spindle, 2-piece construction
Platter Bearing	Hardened 16mm axle bearing
Record Clamp	True reflex style, anodized Aluminum
Motor	Lorenzi 2-pulse, precision 24V DC motor
Tonearm	Dual pivot design bearing
Speeds	33.3rpm, 45rpm, 78rpm
Dimensions (W x H x D)	20.6 x 8 x 12.4 inches / 523 x 203 x 315 mm
Weight	56.4lbs / 25.5kgs
Available Finishes	American Cherry hard wood, Piano Black

Arcam

United Kingdom
www.arcam.co.uk

Arcam first began building sound reproduction equipment in 1972, whilst its founders were still science and engineering students at Cambridge University. Arcam exists to bring the highest fidelity reproduction of music and movies into people's homes. They are committed to engineering products that deliver a level of audio performance that is so convincing and lifelike that it connects you straight to the emotional power of the music. If you love music and movies, then you need to experience the magic that an Arcam-based system can produce.

FMJ CD17

CD Player
£999

As Arcam's most affordable CD player, the CD17 is an ideal introduction to the world of serious music playback. Arcam has been building CD players for more than twenty years. All that experience is brought to bear in the CD17 along with the very latest DAC technologies and hundreds of hours of critical listening. The CD17 offers astonishing musical reproduction from CDs. The all important Digital to Analogue conversion is carried out by the very latest Wolfson 8741 DAC driven by an extremely accurate digital clock and audiophile grade supporting components.

DAC	Wolfson 8741, 24-bit multilevel Delta-Sigma DAC
Frequency Response	0.3Hz – 20kHz
Signal-to-Noise Ratio	109dB
Output Impedance	47 ohms
Minimum Recommended Load	5k ohms
Harmonic Distortion	<0.002%
Power Consumption	32W
Dimensions (W x H x D)	16.9 x 3.3 x 11.4 inches / 430 x 85 x 290 mm
Weight	11.2lbs / 5.1kgs
Available Finishes	Black, Silver

Arcam

United Kingdom
www.arcam.co.uk

Arcam first began building sound reproduction equipment in 1972, whilst its founders were still science and engineering students at Cambridge University. Arcam exists to bring the highest fidelity reproduction of music and movies into people's homes. They are committed to engineering products that deliver a level of audio performance that is so convincing and lifelike that it connects you straight to the emotional power of the music. If you love music and movies, then you need to experience the magic that an Arcam-based system can produce.

Solo Neo

Music System
£1,350

Arcam is proud to introduce the Solo Neo, a music system with superb music performance and network audio capabilities allowing it to play stored music in a multitude of formats along with its internal CD player and radio tuner. With improved audio quality from highly developed pre and power amplifier stages the Solo neo offers a high performance, stylish and easy to use package that delivers music in a way that will thrill any listener. The Solo Neo doesn't sacrifice audio quality for convenience. The internal 75W per channel amplifiers perform superbly and are capable of driving much more demanding speakers in larger rooms.

DAC	Wolfson 24-bit multilevel Delta-Sigma DAC
Frequency Response	20Hz – 20kHz
Input Impedance	47k ohms
Signal-to-Noise Ratio	105dB
Output Power	2 x 75W
Output Impedance	500 ohms
RF Tuning Range	89.5Hz – 108MHz
Harmonic Distortion	<0.3%
Dimensions (W x H x D)	16.9 x 3.1 x 13.7 inches / 430 x 79 x 350 mm
Weight	16.9lbs / 7.7kgs
Available Finishes	Silver

Astell&Kern

Korea
www.astellnkern.com

The Astell&Kern is the portable MQS (Mastering Quality Sound) audio system that brings the original sound, bit to bit, right at the moment of studio recording. MQS refers to all of the high resolution digital music recording studio formats up to 24-bit/ 44Hz – 192kHz. MQS delivers vast amounts of better music experience because it delivers 6.5 times of the detailed information in the same track. Astell&Kern is the way music is meant to be heard!

AK 120

Portable Hifi Audio
$1,399

The finest in Hi-Fi Audio performance in a very compact package! The AK120 not only meets the standards of hi-fi audio, but also utilizes two DACs maximizing the dual-mono set up to completely separate and isolate the left and right audio channels. As a result, the AK120 delivers broader a dynamic range and wider soundstage so that you can enjoy an even more realistic music experience. The red ring wrapped around the volume wheel represents the design statement of the Astell&Kern AK120's identity.

DAC	2 x Wolfson WM8740 (24 bit/192kHz)
Display	2.4 inch IPS Touchscreen (320 x 240 pixels)
Memory	64GB Internal (NAND), 64GB Dual Micro SD Card Slots
Signal-to-Noise Ratio	113dB
Audio Formats Supported	WAV, FLAC, MP3, OGG, APE, WMV, AAC, ALAC, AIFF
Frequency Response	20Hz - 20kHz
Compatibilty	Windows XP, Vista, Windows 7, Windows 8, Mac OS X 10.6.5+
Harmonic Distortion	<0.0005%
Continuous Playback Time	Up to 14 hours (Standard - FLAC)
Dimensions (W x H x D)	2.3 x 3.5 x 0.6 inches / 59 x 89 x 15 mm
Weight	5oz / 0.14kgs

Audiowood

Florida, USA
www.audiowood.com

Audiowood manufactures and sells distinctively styled, sustainably-produced audio gear and other home accessories. Audiowood products include solid wood turntables, energy-efficient amplifiers, home speakers, LED lighting, and other electronic accessories. Audiowood was founded in 2008 by Joel Scilley, a longtime carpenter and woodworker, ex-academic, and designer. Audiowood has ongoing collaborations with Paramount Pictures, SHFT.com, Cornerstone, and Anthropologie.

The Big Easy

Turntable
$1,700

Audiowood makes wooden stereo gear. They started making turntables just about 5 years ago, and have been riding the waves of burlwood turntable mania ever since! Their goal, as always, is to make and sell reasonably priced, stylish, quality stuff that is sustainably-produced. Audiowood's new "flagship" turntable, The Big Easy, is the production version of the one-of-a-kind bachelor. Its features like solid walnut construction with precision CNC-machined, veneered mdf core, the semi-decoupled, solid-walnut armboard with easy, one-bolt mounting for use with multiple tonearms, sets it apart from anything else in its price range.

Platter	Heavyweight glass
Dustcover	Acrylic
Options	Threaded inserts in legs for Rega feet or spikes
Tonearm Length	9 inch, 10 inch
Dimensions (W x H x D)	23 x 7 x 14.5 inches / 584 x 178 x 368 mm
Weight	15lbs / 6.8kgs
Available Finishes	Natural Walnut, Ebonized Black Walnut

Baetis Audio

Montana, USA
www.baetisaudio.com

Today, the very highest audio quality comes not from an ultra-high-end CD/SACD player. It comes from a Windows® PC, specially constructed, using JRiver Media Center® software. And, the computer rips CDs, Blurays, DVDs, DVD-A, etc., and perfectly bit-streams them. If you are an audiophile, find out what you've been missing. Baetis Audio exists for one reason -- to build the very best audiophile computers, at the lowest possible prices.

Revolution II

Media Server
$2,995

Introduced in April, 2012, the Revolution has been upgraded with 3rd generation Core iX CPUs and motherboards. It has the industry's only galvanically-isolated and dedicated SPDIF output. The audio, by actual a-b comparisons, has better accuracy AND musicality than the USB outputs of factory laptops or the custom do-it-yourself computer builds. Their customer support is via telephone and remote desktop control – it has been called "world class" and it is necessary for anyone new to computer audio/video. Baetis also provides the very best in switch-mode, linear, and LiFePO4 power supplies for those who want the best in 2-channel or multi-channel digital audio.

RAM	16G DDR3 1600
SSD	128G Crucial, 512G
Optical Drive	CD/DVD/Blu-Ray
Inputs / Outputs	BNC S/PDIF, Ethernet, USB, HDMI, DVI, VDA, Toslink, Wifi, DC
Software	JRiver (Upgradeable)
Operating System	Windows 7, Windows 8
Processor	Intel Core ix
Options	LiFePO4 Power Supply
Dimensions (W x H x D)	9.5 x 4.3 x 10 inches / 241 x 108 x 254 mm
Weight	8lbs / 3.6kgs
Available Finishes	Black, Silver

Burmester

Germany
www.burmester.de

Since 1977 Burmester has developed and produced hand–made, high–end devices of the absolute top class for the home audio sector in Berlin. Today, they are distributed worldwide. Since the first day, Burmester has moved within a creative field between the priorities of traditional craftsmanship and cutting–edge technology. "Art for the Ear", every single unit is a technical, visual and haptic masterpiece. The distinctive design, finest materials and superb workmanship guarantee ultimate performance and peerless pleasure for all senses.

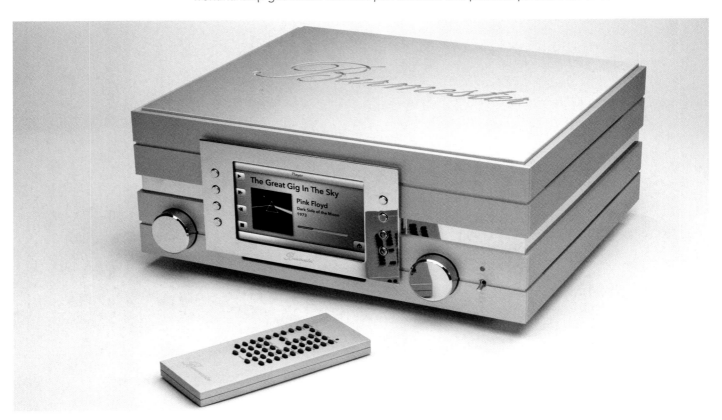

111 Musiccenter
Integrated Music Player
$49,995

The 111 Musiccenter is a real all-rounder, and it can certainly do far more than a server. On account of its fully-fledged preamplifier unit and its integrated reference-level DAC, it can act as the central element of your hifi setup. Three analog and six digital ports are available to the music lover and enable a number of other devices to be attached. The preamplifier section is isolated galvanically from the server section in order to exclude potential interference from the server unit and the analog section with the integrated DAC converter.

Inputs / Outputs	XLR, RCA, TOSLINK, Headphone jack
Capacity	1TB
System Storage	SSD drive
Audio Formats Supported	FLAC, wav, mp3
Sampling Rate	96kHz/24 Bit, 192kHz/24 Bit
Display Panel	High-resolution 7 inch
Dimensions (W x H x D)	17.9 x 8.5 x 15.9 inches / 455 x 215 x 405 mm
Weight	61.7lbs / 28kgs

CasaTunes

Florida, USA
www.casatunes.com

CasaTunes was founded by David Krinker in 2005 to fill a need. David was building a high end home in Florida and needed a high quality music server for his existing whole house music distribution system. Later, he contacted Maynard Knapp, founder of Maynard Electronics, who had a similar need. The two became partners and designed hardware and software for a complete whole home music system and later added wireless capability. CasaTunes has become known for their flexible design and the ease of updating with new features.

CT-12

Music System
$3,499

The CasaTunes CT-12 Music System is part of a family of music systems for small to large homes. The CT-12 supplies 12 stereo pre-amp outputs and 10 or more wireless outputs of CD lossless sound. You can listen to any combination of 12 music streams and 3 external inputs in any room(s). Stream from your choice of internet music services, Windows Media Music or iTunes disk based music stored on any computer in the home. You can even send music wirelessly from an iPad or iPhone through CasaTunes to any combination of rooms. All CasaTunes systems are controlled by optional wall-mounted keypads. That's flexibility!

Network	10/100/1G Ethernet
Frequency Response	20Hz – 20kHz
Signal-to-Noise Ratio	90dB
Harmonic Distortion	<0.05%
Intermodulation Distortion	<0.03%
Output level	3.2Vp-p
External Inputs	3
Dimensions (W x H x D)	15.4 x 2.7 x 13.7 inches / 391 x 69 x 348 mm
Weight	30lbs / 13.6kgs

Canor Audio

Slovakia
www.canor-audio.com

Canor is a new star on the high end tube sky producing the ultimate design "cost no object" electronics. The products incorporate internally the best of classic tube technology. They are super heavy inside with lots of the very best components but outside has a very clean and functional design. Canor has been developing and manufacturing high-end audio products since the eighties. All technological processes in the production lead to the only aim, to manufacture high-quality and reliable products.

CD2 VR+

Compact Disc Player
$3,990

The CANOR CD2 VR + is a new aluminum design tube compact disc player unmatched for its brand new board design. The large display is beautifully readable from standard audible distance. The new product range is characterized by harmonious combination of comfort, workmanship and aesthetics combined with high-quality sound. The unparalleled tube wiring circuit enables to reach natural sound and great detail. The CD2 VR+ compact disc player excels the original concept with plenty of sophisticated features and technical details, which excludes from the crowd of ordinary CD players. Meticulously designed power supply allows to create soft and massive, yet detailed sound, deprived of any interference voltages.

Frequency Response	20Hz – 20kHz
Harmonic Distortion	<0.005%
Signal-to-Noise Ratio	>102dB
Output Impedance	<200 ohms
Output Voltage	2.9V
Power	230V
Dimensions (W x H x D)	17.1 x 4.8 x 14.5 inches / 435 x 122 x 370 mm
Weight	26.4lbs / 12kgs
Available Finishes	Silver, Black

Clearaudio

Germany
www.clearaudio.de

Clearaudio's wide and ever-growing range of products is no accident. Right from the start, back in 1978, they decided to offer a complete range of all the products needed for excellent analog music reproduction. Within their first year they were already the first company worldwide to use boron for their cartridge cantilevers! They haven't stopped innovating since. At Clearaudio, innovation happens at many different levels. Ongoing technological invention. Constantly striving for more intelligent use of materials.

Clearaudio Master Innovation Wood Turntable

Turntable
$30,000

The Clearaudio Master Innovation Wood turntable is the ultimate expression of the Innovation Wood series. It features a belt-driven magnetic platter drive system, which is based upon the award winning and incomparable Clearaudio Statement turntable, and stainless steel sub-platters. The upper platter uses 85mm Ceramic Magnetic Bearing, which allows the platter to float on a layer of air. The tri-star plinths feature a solid Panzerholz (bulletproof wood) and aluminum sandwich construction.

Motor Drive	High torque DC-motor
Speed Control	Electronic Optic Speed Control via infrared sensor
Main Bearing	85mm Clearaudio patented CMB
Platters	POM (anti-resonant), Stainless Steel, Acrylic
Speed Accuracy	<+/-0.05%
Speeds	33.3rpm, 45rpm, 78rpm
Arm Length	9 inch and 2 others
Dimensions (W x H x D)	18.5 x 18 x 19 inches / 479 x 457 x 484 mm
Weight	106lbs / 48kgs
US Distributor	www.musicalsurroundings.com

Horning Hybrid

Denmark
www.horninghybrid.com

Amongst the fine arts, music seems unique in its ability to stir our emotions, whether at live performances, from media broadcasts or in recordings that we all can enjoy at home. The company is particularly proud that Horning has achieved their status amongst knowledgeable enthusiasts without hype, advertising or marketing, they convince on hearing. The performance of their products is such that its demands are just as great; only the finest sources and amplifiers will do it justice.

SATI Transference

Turntable
€26,000*

The Sati Transference ultimate zigma record player is developed for maximum uncolored transference of recorded information in the record groove. To follow this goal means building an uncompromised record player with the best materials you can find, with clever combination of old and new knowledge. The performance of the Sati is like listening to direct studio mastertapes. Each time the speed fluctuates the listener looses vital information that separates reality from 'hi-fi'. Sati narrows this gap like no other turntable in history.

Prices may vary as per color choices.

Platter	Cast Aluminum, 50kgs
Bearing	12mm Tungsten Ball
Motor	Flat high inertia 12 pole
Plinth	40mm thick Aluminum
Arm Board	150mm Aluminum tower
Available Finsishes	Anthracit, 24 carat satin colors

McIntosh Labs

New York, USA
www.mcintoshlabs.com

Founded in 1949 and handcrafted in Binghamton, NY, McIntosh is a global leader in distinguished quality audio products. McIntosh continues to define the ultimate home entertainment experience for consumers around the world, with the iconic "McIntosh Blue" Watt Meters recognized as a symbol of quality audio. Since its inception, McIntosh has been powering some of the most important moments in music history, including Woodstock and the Grateful Dead's legendary "Wall of Sound." McIntosh has not only witnessed history, it has shaped it.

MT5
Turntable
$6,500

The MT5 Precision Turntable maintains classic McIntosh styling by incorporating the iconic handcrafted blackglass front panels on the chassis — which is expertly designed to suppress noise, resonance and coloration for a crisp, clean sound. The illuminated platter's blue-green glow further complements the signature McIntosh aesthetic. Additionally, the MT5 is compatible with any McIntosh stereo preamplifier and integrated amplifier, making it a great piece to complete any two-channel stereo system.

Cartridge Type	Moving Coil
Bearing	Magnetic Sapphire and Ceramic
Arm Tube	Dural-Aluminum
Stylus	Elliptical Diamond
Platter & Bearing	Silicon Acrylic & Magnetic Suspended Ceramic
Frequency Response	20Hz – 50kHz
Channel Separation	35dB
Speed	33rpm, 45rpm, 78rpm
Load Impedance	1000 ohms
Dimensions (W x H x D)	17.5 x 7.5 x 19 inches / 445 x 191 x 483 mm
Weight	31.5lbs / 14.3kgs

McIntosh Labs

New York, USA
www.mcintoshlabs.com

Founded in 1949 and handcrafted in Binghamton, NY, McIntosh is a global leader in distinguished quality audio products. McIntosh continues to define the ultimate home entertainment experience for consumers around the world, with the iconic "McIntosh Blue" Watt Meters recognized as a symbol of quality audio. Since its inception, McIntosh has been powering some of the most important moments in music history, including Woodstock and the Grateful Dead's legendary "Wall of Sound." McIntosh has not only witnessed history, it has shaped it.

McAire

Integrated Audio System
$3,000

The McIntosh McAire™ is a free standing, room filling, wireless sound system, in a sleek, compact unit designed to be placed anywhere you desire. With its distinctive, classic façade featuring the iconic "McIntosh Blue" Output Meters and handcrafted black-glass front panels, the McAire looks as good as it sounds. The single component design eliminates the need for in-wall wiring and professional installation. Multiple McAire systems can be synced to one device, creating a complete in-house audio system effect.

Drivers	2 x 4 inch woofers, 2 x 2 inch Inverted Dome midranges, 2 x 0.75 inch dome tweeter
Frequency Response	60Hz – 45kHz
Crossover Frequency	250Hz; 4.5kHz
Power Requirements	100V, 110V, 120V, 220V, 230V, 240V
Power Consumption (Standby)	<0.5W
Inputs	Wired RJ-45, 10/100 Base-T, Wireless, WPA, Apple AirPlay, USB
Dimensions (W x H x D)	19.4 x 8 x 17 inches / 493 x 203 x 432 mm
Weight	31lbs / 14kgs
Available Finishes	High Gloss Black
Grill	Black knit cloth with magnetic attachment

Oracle Audio

Canada
www.oracle-audio.com

Since 1979, Oracle Audio has been satisfying a passionate desire of thousands of audiophiles in the world. They define themselves as passionate artists who like music, the real, true music. This is the definition of all at Oracle Audio and it is with this highly inspiring and motivational force in mind that they design their products. Oracle Audio products have a signature. The signature of passion and dedication towards all the artists who encapsulate their art in the recorded music they so brilliantly create.

Delphi MK VI

Turntable

£9,345

The Delphi MK VI is Oracle Audio's flagship product. The Delphi has been on the market since 1979. It has been acclaimed worldwide. It is a remarkably handsome work of art. It represents a significant improvement over its predecessors. The new MK VI is engineered to a level that the industry is invited to emulate. Responsible for Oracle's reputation and prestige, the Delphi further demonstrates its ability to defend the leading edge position for another decade, well into the New Millennium. See and hear for yourself. The Oracle Delphi MK VI - a truly outstanding and unique sculpture designed for those who expect more from a record player.

Body	Vented machined
Stylus	Nude line-contact diamond, mirror polished
Coil	Pure iron cross
Output Voltage	2.5mV
Internal Impedance	95 ohms
Frequency Response	20Hz – 20kHz
Channel Separation	Better than 35dB
Tracking Ability	315Hz at a tracking force of 2
Recommended Tonearm Mass	Medium to High
Dimensions (W x H x D)	19 x 6 x 14.5 inches / 475 x 150 x 363 mm
Available Finishes & Weight	Acrylic: 35lbs / 16kgs, Granite: 60lbs / 27kgs

Origin Live

United Kingdom
www.originlive.com

Origin Live is a highly innovative UK based company that designs and manufactures turntables, tonearms, loudspeakers and cables. Founded in 1986, it has built up a wealth of experience leading to numerous awards. With it's renowned musical and entertaining sound quality – reviewers often refer to their products as "addictive", "the type of sonics that makes you keep listening to record after record". Other superlatives reserved for true market leaders are found in numerous reviews.

Sovereign MK3

Turntable
£4,770

The Sovereign MK3 turntable uniquely employs a single point, very stiff suspension. Although the arm and platter are effectively supported and "earthed through a single point, the deck is simple to set up and behaves like a non-suspended design. Other features include a high grade dc motor, extensive decoupling of main bearing and arm board. Now in MK3 version, this deck is a highly acclaimed award winner and has been reviewed as close to the ultimate in performance and style available today. Sound is characterised as powerful and dynamic with superb transparency and tangible sound staging.

Speed	33rpm, 45rpm
Speed Accuracy	0.05%
Arm Length	9.5 inches
Options	12 inch tonearm, Dual armboard with deck
Dimensions (W x H x D)	17.7 x 5.9 x 14.9 inches / 450 x 150 x 378 mm
Weight	63.9lbs / 29kgs

Pro-Ject
Audio Systems

Austria
www.project-audio.com

In 1990, Pro-Ject Audio Systems was founded by Heinz Lichtenegger, in Vienna, with the idea that analog playback is the most cost-effective way to listen to music of audiophile quality. Inspired by the concept of supporting analog in the face of the digital onslaught, Pro-Ject became one of the most powerful driving forces in reinventing analog turntables for the hi-fi market. Today, Pro-Ject is the world leader in the manufacture of quality hi-fi turntables, as turntables are once again accepted as a must-have for every concerned music lover.

Debut Carbon

Turntable
€29,900

The Debut Carbon sets a new standard in terms of cost versus performance by evolving the original Debut to a higher level of build precision, sound quality and aesthetics. Pro-Ject has once again challenged convention by offering even better performance at an easily affordable price. The most profound upgrade is the addition of the 8.6" single-piece tube carbon fiber tonearm that increases stiffness while decreasing unwanted resonances resulting in a higher fidelity presentation of your treasured recordings.

Speed	33rpm, 45rpm
Drive Principle	Belt Drive
Platter	300mm metal with felt mat
Mains Bearing	Stainless Steel
Tonearm	8.6 inch Carbon
Signal-to-Noise Ratio	-68dB
Arm Length	218.5mm
Dimensions (W x H x D)	16.3 x 0.7 x 12.5 inches / 415 x 18 x 320 mm
Weight	12.3lbs / 5.6kgs
Available Finishes	Black, Blue, Red, Green, Yellow, Grey, White

PTP Audio

Amsterdam
www.ptpaudio.com

PTP Audio is the company of Peter Reinders and is founded on years of experience. To ensure that the product you buy is as good as it can be, Peter Reinders services all Lenco parts and assembles, tests and tunes each individual turntable himself. All new parts are made, according to their design and specifications, by highly skilled manufacturers to guarantee a long, hassle free life and a perfect fit and finish. PTP Audio turntables unite the best of yesterday with the best of today to create a unique product capable of the finest music reproduction.

Solid 12

Turntables
€2,650

Listening to an idler drive turntable is not at all like hearing a belt drive model. Idler drive turntables let the music come first. You will find yourself tapping your foot instead of focussing on sound quality. Idlers simply have incredible musicality and energy. All Solid 12 turntables are made to order, available in a wide range of colours, and assembled by hand. The result is a turntable that has the sonic signature of an idler drive, but with the looks and reliability expected today. PTP Audio turntables focus on music, a quality far rarer than it ought to be.

Design	Idler drive
Top Plate	Two part 4mm stainless steel
Features	Separate motor plate for maximum isolation
Platter	Polished 4kg die cast Aluminum
Motor	High torque AC induction motor
Speed	33rpm, 45rpm
Arm Length	12 inch arms for 270mm – 330mm
Dimensions (W x H x D)	21.6 x 1.9 x 17.7 inches / 550 x 50 x 450 mm (Corian Plinth)
Weight	55lbs / 25kgs
Available Finishes	Black, White & Others

Roksan Audio

United Kingdom
www.roksan.co.uk

Founded in 1985, Roksan Audio has since developed into one of the best known and highly regarded names in audio. The company holds numerous worldwide industry awards and honors, making them continually one of the most successful companies in specialist high-level audio reproduction. Roksan's initial success was gained with the breakthrough Xerxes record player and progressed with a host of designs –from CD players and amplifiers through to streaming devices, that have become classics in the audio world.

Oxygene
CD Player
$4,312

Oxygene CD Player is the end result of countless hours of experimentation, fine-tuning and a relentless dedication to sonic perfection. Immediately, you will notice the lack of buttons and switches. A sleek, innovative design with only the words 'LESS IS MORE' discretely printed on top of the cabinet. These three touch sensitive words control all functions of the player. Touch MORE to select next track on the disc or LESS to select the previous track. Touching IS gives access to more functions. Along with the clean, modern aesthetics and you have a highly desirable product line that is not only a joy to own but a joy to use.

DAC	24 bit/192kHz - BB PCM1796
Analog Output	Single ended via RCA
Digital Outputs	Coaxial, Optical
Signal-to-Noise Ratio	<100dB
Harmonic Distortion	<0.002%
Power Consumption (Idle)	8.2W
Mains Power	100V / 120V / 230V @ 50Hz
Maximum Output	2.3Vrms
Dimensions (W x H x D)	12 x 2.3 x 12 inches / 305 x 58 x 305 mm
Weight	9lbs / 4kgs
Available Finishes	Black, White, Silver, Gold

Rotel

Japan
www.rotel.com

Rotel was founded in 1961 and is still under the same family ownership. They manufacture a full range of high performance audio/video receivers, surround sound processors, power amplifiers, CD players and digital audio components in their own factory. Rotel's latest models combine state-of-the-art digital and classic analog circuits to deliver surpassing fidelity.

RCD1570

CD Player

$1,000

The culmination of more than three decades of Rotel CD technology refinement, the RCD-1570 is the flagship CD player in Rotel's new 15 Series, combining a wealth of features to reveal even the subtlest nuances in your CDs. The CD player's Wolfson WM8740 digital filter/stereo DAC ensures reference levels of reproduction under all circumstances. The RCD1570's slot-loading disc transport design isolates the CD itself from potentially destructive vibration modes that may affect musical definition, while simplifying the front panel's appearance.

Frequency Response	20Hz - 20kHz
Distortion	0.0045%
Signal-to-Noise Ratio	>100dB
Dynamic range	>96dB
Power Consumption (Standby)	15W
Channel Separation	>98db@1kHz
Channel Balance	+/-0.5db
Output Impedance	100 ohms (RCA); 200 ohms (XLR)
Dimensions (W x H x D)	17 x 4 x 12.6 inches / 431 x 93 x 320 mm
Weight	14.74lbs / 6.7kgs

Soulution

Switzerland
www.soulution-audio.com

Purity of sound. No adulteration. No artificial enhancement. Nothing added. Nothing left out. It's a goal many audio manufacturers aspire to. But few attain. Soulution does. Without marketing gloss, just quietly understated design on the outside and cost-no-object technology inside, the result is audio for a truly discerning audience. Simply a uniquely sensual, organic experience that delivers all the honesty, the power, the subtlety and the emotion of music. Just as nature intended.

Soulution 540
Digital Player
$32,000

Not even the best amplifier or loudspeaker can compensate for what has been lost at the source. That is why the quality of a recording, and of the device used for its playback is so fundamental to high fidelity. With the Soulution 540 digital player, you lose nothing of the original recording. Simply, it reads all the musical information and transfers it into the analog world. Purely. Authentically. Perfectly. Beautifully.

Output Voltage	Balanced: 4Vrms; Unbalanced: 2Vrms
Peak Output Current	0.2A
Output Impedance	Balanced: 10 ohms; Unbalanced: 10 ohms
Frequency Response	DC - 100kHz
Harmonic Distortion	<0.002%
Noise Floor	140dB
Digital-In Sensitivity	0.3V - 5Vp-p
Output Stage Bandwidth	20MHz
Power Consumption (Standby)	<0.5W
Dimensions (W x H x D)	17.6 x 5.62 x 17.4 inches / 448 x 143 x 442 mm

Symbol

New York, USA
www.symbolaudio.com

Symbol Audio handcrafts modern audio Hi-Fi consoles and vinyl LP storage cabinets in the tradition of fine furniture. Their consoles bring together time honored analog electronics with digital technology to deliver audiophile sound, enriching the connection between the listener and their music. The collection is designed to provide distinctive audio products for the design and quality oriented consumer, and is a singular alternative to impersonal mass produced consumer electronics and storage cabinet options available on the market currently.

Stereo Console
Media Console Table
$3,000

The Stereo Console brings your music to life, no matter how you play it. Connect up to 3 devices to the built-in amplifier and enjoy the impressive soundstage while streaming audio via Bluetooth, spinning records, or watching TV. Modular storage for LP's and/or components lets you customize your console to fit your lifestyle. Each LP bin holds up to 100 LP's; hang up to 3 LP bins on the steel base. Component storage bins accommodate most A/V components like a DVD player or receiver box.

Amplifier	High Efficiency 2.1 channel Class D
Drivers	4 inch full range titanium cone, 8 inch subwoofer
Compatibility	Any Audio Source -Turntable, iPod, TV, Bluetooth
Inputs	RCA, 3.5mm
Crossover	110Hz
Features	Removable back panel for component storage and wire management
Options	LP storage bins, AV component storage
Table Heights	33 inch console table, 24 inch media console
Dimensions (W x D)	48 x 15.7 inches / 1219 x 399 mm
Available Finishes	Natural Walnut, Natural Maple with Black, White

Synthesis

Italy
www.synthesis.co.it

Synthesis is a dynamic, fast growing company that looks to the future and, compared with other companies in this field, has been able to anticipate the metamorphosis of the various success factors in the Hi-Fi market. From its foundation in 1992, the "Synthesis Art in Music ®" has had as its main objective-the design of equipment whose ultimate aim is not just sound reproduction, but also to offer a product that represents refined elegance and musicality all rolled into a single element.

Pride
CD Player
$2,200

These things of beauty are not arrived at by accident, all of their flowing features are bound by the one outcome-their design. This unique combination of materials both natural and man-made are brought together not just as a product, but as something to be treasured. This interpretation of hand crafted design is achieved in a way only the Italian's sense of flair and taste knows how to give. Shapes, materials, finishes, technologies, all joined by a constant thought: designs and sounds that enriches your quality of life. If together a well being can be brought by their designs then... feel good.

Frequency Response	20Hz – 20kHz
Signal-to-Noise Ratio	>95dB
Harmonic Distortion	<0.1%
Digital Conversion	192kHz / 24 Bit
Output Level	2V
Power Consumption	25W
Front Controls	Standby, Open, Play, Pause, Stop, Next, Previous, Display, IR receiver
Options	Remote Control
Dimensions (W x H x D)	12.5 x 2.3 x 8.6 inches / 320 x 60 x 220 mm
Weight	11lbs / 5kgs
Available Finishes	Ecologically sensitive lacquer

Acoustic Preference

Slovenia
www.acoustic-preference.com

Acoustic Preference enterprise is a small family-run manufacturing company established in 2001. With the help of music, we can conjure up the majority of human feelings. At the same time, music represents a tool for relaxation and creativity, since it influences so many human senses. Acoustic Preference's goal is to create acoustic components that combine natural sounds and offer great delight when listening to music on every occasion. They want to present their clients with durability and a feeling of prestige and excellence.

Verismo

Interconnect Cable
$825 (1m pair)

As with every chain system, the weakest link in your audio system will be your sonic limit. Verismo interconnect is the company's newest product. With its design and execution it represents and reflects the main goal of their firm, which is neutral but flawless reproduction of audio signal on every level. Verismo is balanced, directional and a double shielded audio cable. Materials used were selected very carefully, according to their properties represented in audio field. Silver, Teflon and Cotton. And finally, each interconnect is assembled by hand, from first piece of wire to last layer of dielectric and termination.

Conductor	22AWG pure solid silver
Insulators	Teflon, Cotton
Shielding	OFC copper braid and OFC copper foil
Terminals	Copper Rhodium RCA
Standard Length	1m

Art Vinyl

United Kingdom
www.artvinyl.com

Founded in 2005, Art Vinyl is the original way to display your favourite vinyl records; perfect for music or art enthusiasts and those with an eye for stylish interior design. The first real champions of Art & Design for vinyl records, Art Vinyl also curate exhibitions celebrating the best in sleeve design. Art Vinyl record frames give you the chance to stylishly display your vinyl albums, 12" records, and their contents on your wall and change the display without removing the frame from the wall.

Play & Display

Album Flip Frame
$159.99 (Triple Pack)

The Play & Display Flip Frame was invented by founder Andrew Heeps, then developed by award winning design agency Webb Scarlett DeVlam. The product is so original and innovative it has a patent already granted in 4 countries with many more to follow. This product is a result of ingenious design from some of the best product designers in the market combined with a team so passionate about Art and Music they had to bring Play & Display to market. And so, got the top designers to make it happen.

Features	Cushioned back plate, Wall mounting
Options	Single Frames available
Size	12 inches
Available Finishes	Black, White

Atlas Cables

Scotland
www.atlascables.com

Atlas is a successful, independent, UK-owned specialist cable company based in Kilmarnock, Scotland. They engineer their products for performance, so everything Atlas does is based on extensive research and proven test results. This ensures that their products are completely consistent in manufacture and ensures a defined progression as you move up the ranges. In other words, they not only know exactly what they're doing, but they also know why.

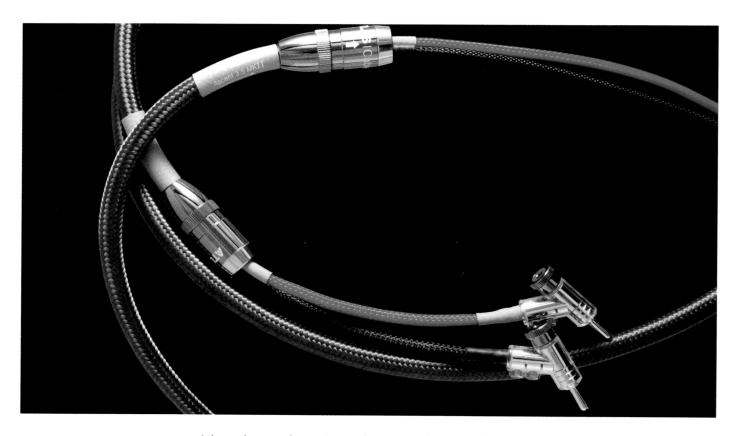

Ascent 3.5 MKII

Interconnect & Speaker Cable
£795 (2m pair)

Atlas only uses the purist quality copper in its products, with the Ascent being no exception. Utilizing Atlas' preferred copper technology, where each strand is 'pulled' from a single copper crystal enabling electrical signals to flow with the fewest deletions, Ascent 3.5 MKII uses 3.5 sq mm cross section of the finest stranded copper to create its conductor. Atlas, with its suppliers, has developed a way of coating conductors with a type of Teflon that mitigates the velocity of signal issues that cheaper dielectric materials exhibit.

Conductor Material	OCC Copper
Dielectric	Teflon
Termination	Expanding Rhodium, OCC spade
Resistance	0.0058 ohms
Capacitance	100.69 pF
Inductance	0.539 microH
OD	9.8 mm

Kharma

Netherlands
www.kharma.com

Kharma embodies the seeking to ultimate beauty, where Kharma converges audiophile excellence with aesthetic beauty. A genuine approach to reveal the most subtle musical experience and to unlock immeasurable aesthetic joyful inner experiences in the listener. Like a piece of art in painting can evoke feelings of beauty and excitement in the viewer, the Kharma products are pieces of art in audio vibrations and they will evoke the highest feelings of beauty and wonder in the listener.

Enigma Extreme Signature KLC

Loudspeaker Cable
$30,000 (pair of 2m)

The KLC-EEXS-1a is the ultra high-end loudspeaker cable based on mono-crystal silver and various diameter pure silver-gold strands. The conductors are treated with the Kharma Advanced Core Treatment, and equipped with the new Kharma Proprietary Insulation, which is combined with air insulation. The conductors are placed behind ultra effective shielding, and mechanical vibrations are minimized. All this leads to a crystal clear and very powerful cable, with extreme revelation of the smallest detail, infinite resolution, enormous dynamics, sweetness, air and a very realistic and transparent, holographic soundstage with an overwhelming musicality.

Type	KLC-EEXS-1a
Options	Available as Bi-wire and Jumper cables
Connection types	Spade, Banana
Minimum Length	1.5m
Standard Length	2m

KingRex Technology

Taiwan

www.kingrex.com

KingRex, provides high-end cables. Starting in the PC HiFi area, KingRex is a trend leading designer for digital stream devices. After they launched their first DAC with USB—T20U, they established a creative image in the audio industry. Along with providing high-end solutions, all of KingRex products are made to be high performance and a reasonable price. Their mission is to please your ear! Enrich your life! They are innovative by desire!

Unanimous U-Craft (Y)

USB Cable

$599

The Unanimous Series USB cable has two types, one is Y shape, and the other one is S shape. It is made for decreasing the effects from PC substantially. No loss or no distortion is encountered, due to its special design and treatment, while signal transmission. The flat design, where signal and power are parallel instead of binding together reduces the distortion. Cryogenic treatment is used in -196 centigrade, to make the element firmer, steadier, which in turn improves the signal transmission. The Y type has two A plugs and one B plug, one A is for PC signal, the other one is for external 5V power supply.

Conductor	Silver-plated OFC
Cable Design	4 signal Conductor Symmetricon TM
Structure	Flat/Inverted Concentric Structure
USB Plug Type	Y type – A,B
USB Plug	1 micro inch 24k gold plated
USB Plug Case	Aluminum CNC in red anodize with special damping
Length	2 meters
Available Finishes	Blue

LessLoss

Lithuania
www.lessloss.com

LessLoss is a high performance audio company. Since its internet inception five years ago, Less-Loss has shipped over 3000 high performance products to a worldwide customer base in 65 countries. LessLoss products have received four press awards and over 200 reviews. Technologies include Skin-filtering, the use of Panzerholz to minimize micro-vibration, black body ambient field conditioning, as well as a breakthrough distortionless analog signal transfer technology called Tunnelbridge.

DFPC

Power Cable
$1,149 (2m pair)

The DFPCs (Dynamic Filtering Power Cables) are much more than mere power cords: each model employs a different grade of LessLoss Skin-filtering, representing not only substantial leaps in performance, but the most elegant and effective power filtering technology ever manifested in a power cord. LessLoss's Skin-filtering technology prevents undesirable noise from entering your equipment—without impeding dynamics. With each grade of Skin-filtering, the DFPCs provide progressively clearer insight into your signal's character, guaranteeing an audibly more faithful presentation of the music.

Technology	Skin-Filtering
Design	Robust
Construction	Braided
Resistance	Low
Current Rating	41A – 60A
Plug & IEC Options	P-079, Schuko / 15A, 20A
Flexibility	High
Connector	Oyaide 079
Live-wire Cross Section	Original: 6mm2, Signature: 12mm2, Reference: 12mm2
Available Lengths	1m, 1.5m, 2m, 2.5m, 3m
Available Finishes	3 types of designs

MG Audio Designs

Colorado, USA
www.mgaudiodesign.com

MG Audio Design, LLC, was formalized as a company about 2 years ago but has been 25 years in the making. Their founding purpose is to offer state-of-the-art audio components at costs significantly less than most state-of-the-art equipments on the market today. To date, the company has focused on speaker wires and interconnects, but they continually examine opportunities to improve other products in the equipment chain as their experience and expertise permits.

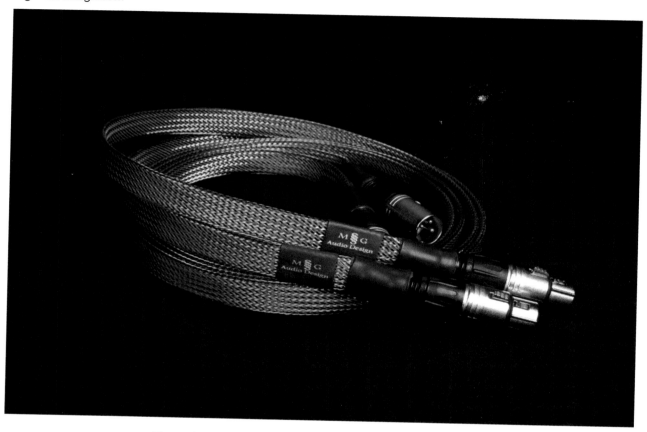

Planus AG

Interconnect
$1,100 (1m pair)

Planus AG is MG Audio Design's state-of-the-art interconnect. AG is magical in its musical presentation and is likely the most detailed, transparent wire on the market today. Its ability to resolve low level detail and musical subtleties is stunning. You can easily hear the bowing and plucking of strings, sticks hitting drum-heads, hammers striking the strings of a piano and the subtle voice inflections of a singer. On well-recorded live venues you hear the ambiance of the recording environment and the subtle noises in the crowd. This interconnect can put you in the performance.

Conductor	Silver
Insulator	Teflon
Exterior	TechFlex
Terminations	Silver Bananas, Silver Spades
Versions	Balanced, Single-Ended
Options	Custom lengths available
Lengths	1m, 1.5m

MG Audio Designs

Colorado, USA
www.mgaudiodesign.com

MG Audio Design, LLC, was formalized as a company about 2 years ago but has been 25 years in the making. Their founding purpose is to offer state-of-the-art audio components at costs significantly less than most state-of-the-art equipments on the market today. To date, the company has focused on speaker wires and interconnects, but they continually examine opportunities to improve other products in the equipment chain as their experience and expertise permits.

Planus III

Speaker Cable
$1,900 (8 foot pair)

Planus III is one of MG Audio Design's Reference Series of wires and provides state-of-the-art performance. Planus III is a wire that outperforms other reference level wires in terms of speed, transparency, detail, sound-staging, and sheer musicality. The sound-stage it casts provides a palpable presence to the performers. Planus III pushes the state-of-the-art in speaker wire performance enabling the subtle nuances of the musical performance to be easily heard and the emotional content felt.

Conductor	Copper
Insulator	Teflon
Exterior	TechFlex
Terminations	Silver / Rhodium - Bananas, Spades
Versions	Balanced, Single-Ended
Lengths	6 foot, 8 foot, 10 foot
Options	Custom lengths available

Nordost

Massachusetts, USA
www.nordost.com

Nordost is the premier manufacturer of hi-fi audio cables and accessories in the consumer electronics industry. Nordost takes pride in providing high quality, purpose-built equipment, manufactured in the U.S. for audiophiles and music enthusiasts throughout the global market. The innovative, proprietary technologies employed in Nordost's products make it possible for any sound system to achieve its full potential. Nordost cables provide an unfiltered sound, giving customers the phonic experience of a live show in the comfort of their own, personal soundstage.

Valhalla Reference Cable

Speaker Cable
$5,599 (1m pair)

The Valhalla 2 Reference Cable range is a technologically sophisticated line of cables that delivers an unparalleled performance. The V2 range consists of analog, digital, and tone arm interconnects, as well as loud speaker and power cables. V2 Cables use Dual Mono-Filament technology to achieve minimal insulation contact and excellent mechanical dampening. Additionally, the conductors in the Valhalla 2 range are made with pure, solid core copper, plated with 85 Microns of silver and completed with Nordost's new HOLO:PLUGTM, a patent pending connector designed to transfer every last nuance of detail with maximum efficiency.

Conductor	40 x optimized diameter in micro mono-filament construction
Insulator	High purity class 1 extruded FEP
Material	Silver over OFC solid core
Capacitance	11.8pF/ft
Inductance	9.6 microH/ft
DC Resistance	2.6 ohms/1000ft
Propagation Delay	96% of speed of light
Connectors Available	Spade, Banana
Dimensions (W x T)	2.2 x 0.04 inches / 55 x 1 mm

Purist
Audio Design

Texas, USA

www.puristaudiodesign.com

Purist Audio Design's extensive research and testing gets down to the heart and soul of a product. At this most basic, yet extremely crucial, level is the raw materials. The first key is the choice of precious metals and alloys to be used for the conductors. The company look at not only capacitance, inductance, and resistance, but also magnetic fields. That is why they use alloys for their conductors, rather than simple copper or silver alone, for the ultimate in sonic quality and performance.

Purist 25th
Anniversary
AC Power Cable

Power Cords

$9,900 (1m pair)

The 25th Anniversary AC Power Cable uses power conditioning circuit technology that provides EMI/RFI noise control, filtering, and reduction. It integrates the use of a multistage power conditioning circuitry that passively controls, filters and reduces the effects of EMI/RFI from coupling via direct conduction, induction and capacitive to a level that allows the dynamics of music to breath and prevail. All 25th Anniversary cables come packaged in a handsome soft case.

Conductors	204/60 stranding
Metals	Silver
Shielding	Ferox 103 plated silver braid 98% coverage
Insulator	Thermoplastic Elastmer
Gauge	8 AWG
Current Capacity	75A
Voltage Rating	600V
Resistance	1.07m ohms
Material Treatment	5x Cryomag
Cable Diameter	1.25 inches

Skogrand Cables

Norway

www.skograndcables.com

Skogrand Cables is a small company located in the mountain hills of Norway. In this serene and tranquil environment they manufacture audio cables with the support of several key branches of the Norwegian government due to their product's market leading signal transferring capabilities. You will find Skogrand cables in the reference setups of reviewers and dedicated audiophiles all over the world. Spearheading signal transfer efficiency-delivering pure audio joy.

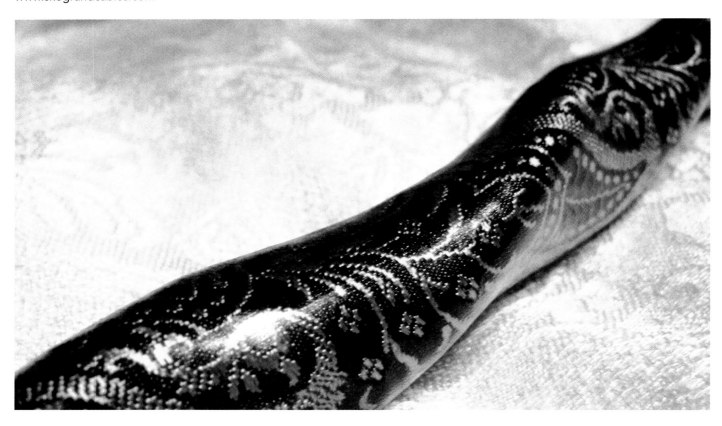

SC Markarian 421

Speaker Cable

$8,150 (2m pair)

The SC Markarian 421 offers solid core copper leads suspended in air within a framework of ultra low dielectric fabrics and PFA tubing. With this cable build, Skogrand has achieved an effective dielectric constant of 1.0018 with a signal transfer speed of 99.82 % of the speed of light. Skogrand Markarian 421 is a marvel of sight and sound delivering pure audio joy to the global audiophile community. What makes these cables truly special according to reviewers around the world and customers is that they give an utterly precise and truthful reproduction of what any component has to offer at its best.

Capacitance	4.57pF
Inductance	0.22 microH – 0.25 microH
Resistance	0.00575 ohms
Attenuation	0.001dB
Conductors	12 AWG pure solid copper leads
Insulator	Air
Framework	Ultra low dielectric fabrics and PFA tube suspension
Construction & Cable Ends	Dual Lead / Balsa framework
Connectors	Banana, Spades, XLR, Speakon
Diameter	0.076 inches / 1.954 mm
Available Finishes	Different Silk brocade cable sleeves

Surreal Sound

Virginia, USA
www.surreal-sound.net

Surreal Sound Audio, based in Chesterfield, VA is home to a new line of hand-crafted luxury speakers. You may never have heard bass as well defined and dynamic as with their full range speaker system. The midrange presentation is smooth and realistic, soul soothing as music should be. The high frequencies when properly done as by them, locate and stabilize the soundstage placing instruments in a virtual 3D world before you. You will always have concert quality sound without ever having to leave the comfort of your home.

Morph Cord

Power Conditioning Cord

$799

The Morph Cord is now an essential component to every system. Meticulously designed to eliminate all hash from electrical circuits powering everything in your system. When used through-out to power everything from amps to DACs the improvement in your system is nothing short of astonishing. The conditioner is proprietary and is able to block RF and EMF better than any other technology. This is the most cost effective improvement for your system without spending thousands of dollars.

Design	Quad Star
Power Conditioner	Built-in
Conductors	12 gauge
Features	Dual quadruple device built-in to eliminate noise andi nterference
Configurations	1 male/1 female, 1male/2 females

Whitworth Design

California, USA
www.whitworthdesign.com

Whitworth Design is a design and fabrication company that strives to deliver exceptional products. They have a diversified portfolio, use cutting-edge manufacturing techniques, and utilize the best materials available. This allows them to offer innovative designs that are second to none. All of their products are manufactured in the U.S.A.. They maintain a quality control level that most companies cannot offer. Their product lines have been tested and given positive results by some of the most respected firms in the world

Pulse Series 2

Speaker Stands
$1,599

The "pulse" series 2 speaker stands are designed for medium to large bookshelf speakers, the aluminum components are machined from 6061 heat treated aluminum alloy, the 3" main vertical tube is fillable for damping with lead shot or silica. These stands feature two 1.5" tubes that add aesthetically and serve as wire management tubes. It has a top platform with an hourglass shape, and the bottom H shaped platform with 4 point legs for soft surfaces. Pucks with silicon pads are available for hard surfaces such as wood or concrete.

Top Platform Dimensions (W x T x D)	7 x 0.4 x 11 inch / 178 x 10 x 279 mm
Bottom Dimensions (W x T x D)	16 x 0.375 x 18 inch / 406 x 10 x 457 mm
Height	29 inch / 737 mm
Available Finishes	Severa anodized colors.

Wireworld Cable Tech.

Florida, USA
www.wireworldcable.com

A perfect cable would perform as if the components were connected directly, with no cable at all. Realizing the power of that essential truth, Wireworld founder and designer David Salz turned it into a development methodology that produced some of the most highly regarded audio cables in the industry. Acclaimed for numerous award-winning and patented innovations, including the world's first high-end HDMI cables, Wireworld has a reputation for exceeding the expectations of audio connoisseurs worldwide.

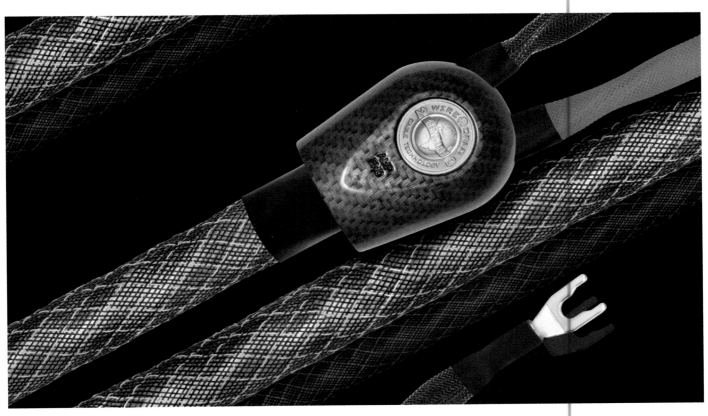

Platinum Eclipse 7
Speaker Cable
$20,400 (2.5m pair)

Wireworld's flagship Platinum Eclipse 7 is the culmination of over three decades of continuous research dedicated to creating cables that preserve every nuance of live musical sound. This cable makes such profound sonic improvements that average recordings take on much of the natural acoustic presence associated with the best audiophile-grade recordings, and the best recordings simply come to life. The scientifically engineered DNA Helix design combined with Wireworld's proprietary Composilex 2 insulation technology and the purest Ohno Continuous Cast® solid silver conductors make this speaker cable the finest in the world.

Conductor	Ohno Continuous Cast solid silver
Gauge	9ga./6mm2, 12ga./3mm2
Insulator	Composilex 2
Design	DNA Helix®
Standard Lengths	2m / 6.5ft, 2.5m / 8ft, 3m / 10ft, 5m / 16.5ft, 6m / 20ft
Options	Custom lengths available, available in standard or bi-wire pairs
Available Connectors	Uni-term interchangeable silver-clad OFC spades and/or bananas

Wireworld Cable Tech.

Florida, USA
www.wireworldcable.com

A perfect cable would perform as if the components were connected directly, with no cable at all. Realizing the power of that essential truth, Wireworld founder and designer David Salz turned it into a development methodology that produced some of the most highly regarded audio cables in the industry. Acclaimed for numerous award-winning and patented innovations, including the world's first high-end HDMI cables, Wireworld has a reputation for exceeding the expectations of audio connoisseurs worldwide.

Platinum Electra

Power Conditioning Cord
$1,700 (1m pair)

The Platinum Electra Power Conditioning Cord exemplifies Wireworld's commitment to producing the world's finest AV cables, regardless of cost. With conductors and plugs made of the ultimate conductor material, Ohno Continuous Cast (OCC) solid silver, and the ultra-quiet Composilex 2 insulation, the material quality alone sets a new standard for the industry. Moreover, the proprietary Fluxfield conductor geometry maximizes filtering to produce the most impressive sound quality improvements ever provided by a power cord.

Conductor	Ohno Continuous Cast solid silver
Gauge	9ga./6mm2, 12ga./3mm2
Insulation	Composilex 2
Design	Fluxfield Technology
Contacts	Ohno Continuous Cast silver-clad copper
Standard lengths	1m / 40in, 1.5m / 5ft, 2m / 6.5ft, 3m / 10ft, 6m / 20ft
Options	Custom lengths available, available in standard or bi-wire pairs
Available Connectors	US, UK, Schuko plugs